The Lost Works of
ISAMBARD KINGDOM BRUNEL

John Christopher

AMBERLEY

Although this book focuses primarily on the lost works of Isambard Kingdom Brunel, and by definition these have either been replaced by later structures or not survived at all, there are still many examples, such as the Hungerford, Chepstow and Clifton bridges, the remains of the Cornish viaducts or the major railway stations, all of which are well worth visiting. But please note that some of the sites covered are on private property and, especially when it comes to the railways, access might also be restricted for reasons of safety.

Opposite: Robert Howlett's famous photograph of I. K. Brunel, taken at Millwall in 1857 during the construction of the *Great Eastern.*

First published 2011

Amberley Publishing
The Hill, Stroud
Gloucestershire, GL5 4 EP

www.amberley-books.com

Copyright © John Christopher, 2011

The right of John Christopher to be identified as the Author of this work has been asserted in accordance with the Copyrights, Designs and Patents Act 1988.

ISBN 978 1 4456 0090 1

British Library Cataloguing in Publication Data. A catalogue record for this book is available from the British Library.

Typesetting by Amberley Publishing.
Printed in the UK.

Left uncompleted for many years, construction work on the Clifton Suspension Bridge at Bristol only resumed in 1863 as a tribute to Brunel who had died in 1859. This photograph from the Clifton side shows preparations for the installation of the chains. Below, the *Great Eastern* at Hearts Content, Trinity Bay, in July 1866. Built at Millwall, sideways to the Thames, the ship wouldn't budge at first and hydraulic ramps were brought in to slowly push her into the river. (CMcC)

Introduction – The Big Picture

Robert Howlett's iconic image of the great engineer, propped against a backdrop of improbably massive iron chains, is widely celebrated as one the most famous photographs of the nineteenth century and, possibly, of all time. Some commentators have even described it as the first 'modern' photographic portrait. Without doubt, it is better known and more instantly recognised than any individual picture of the queen who gave her name to the Victorian era. In the 150 years or so since this photograph was taken it has appeared on acres of book covers and has been re-produced *ad nauseam* on t-shirts, postcards, jigsaws and just about every other item of merchandise you can possibly imagine. As a result, even if you know next to nothing about the life or work of Isambard Kingdom Brunel, you will be familiar with this image.

So how did this come about? How did a photograph of a long-dead civil engineer take on such a potent life of its own to the extent, I would argue, that it almost single-handedly propelled Brunel to the top of the Great Britons poll where he came second only to Winston Churchill? The answer lies in the picture itself and every picture tells a story. Scratch beneath the surface of this one and it speaks volumes. In 1857 Robert Howlett had been commissioned by *The Illustrated Times* to take a series of photographs showing the construction of the *Great Eastern* steamship at the shipyard on the Thames at Millwall. Ironically they were to be used by the newspaper's engravers to create woodblock illustrations as the techniques for reproducing photographs in print had yet to be perfected. While he was at Millwall, taking his pictures of the massive hull which rose like a wall of iron above the surrounding buildings, Howlett came across the ship's engineer and invited him to pose for a photograph. Not particularly concerned about the recording of his own image, Brunel asked a colleague to join him, but he declined. As he later wrote, 'I alone am hung in chains', which is why we have a solitary figure shown leaning with one arm resting against the chains on the winch drum to prevent movement during the long exposure time.

This is actually one of three 'chain' pictures taken of Brunel that day, or at around that time. Another shows him in a more casual pose with the stove-pipe hat at a decidedly rakish angle, and in the third he is seated on the wooden block to the right. However, neither of these conveys the same strength of purpose which we read into the famous photograph, to the extent that it has become a mirror to our own expectations of history. It fulfils and confirms the clichés of the 'Little

Giant' – a cigar clamped in his mouth – as an heroic figure, a man of iron. Now contrast this picture with the very last photograph of Brunel, taken on the deck of the *Great Eastern* on 5 September 1859, around two years later – *see page 101*. By this time Brunel was in poor health, suffering from an incurable kidney condition, and he is shown resting on a walking stick as he poses in front of one of the funnels shortly before the ship was due to head out of the Thames. He is a shadow of his former self and the years of work, and in particular the stress of building and launching the *Great Eastern*, have clearly taken their toll. Moments after this photograph had been taken Brunel collapsed, possibly from a heart attack, and he was carried off the ship and taken to his home. Now, just imagine the difference it would have made to our perception of Brunel, our image if you like, if this final photograph had been the only one we had of him.

It is said that after his collapse Brunel spent the few remaining days putting his affairs in order, but it must also have been a time to reflect upon his life's work. Yes there had been great triumphs – such as the Great Western Railway or the Royal Albert Bridge at Saltash – incredible feats of engineering that have shaped our modern world, but equally there were many magnificent failures. These were the price of genius and the consequence of innovation.

The roll call of lost works is surprisingly long. At the time of Brunel's death his father's great project, the Thames Tunnel linking Wapping and Rotherhithe, was only accessible to pedestrians and had become the haunt of prostitutes and thieves. As for the younger Brunel, his first engineering commission – his 'first love' he called it – the Clifton Suspension Bridge, had run out of money and the forlorn stumps of its towers stood like tombstones over the Avon Gorge at Bristol. In London the Hungerford Bridge was to be demolished with only its support piers left to create a railway crossing into Charing Cross. The Broad Gauge was another lost cause, a bitter blow, condemned forever by the Railway Regulation (Gauge) Act of 1846 which had made the so-called 'standard' gauge mandatory. Then there was the atmospheric traction experiment in South Devon, a humiliating and very public failure that he would rather forget.

Brunel's great ships had fared little better. Following the stranding of the *Great Britain* at Dundrum Bay in 1846, his vision for a transatlantic service running all the way from London via train and ship to New York had been scuppered. Consequently, both of the Great Western Steamship Company's vessels were sold for a pittance. After twenty years in service, the wooden-hulled *Great Western* paddle steamer was scrapped at Castles Yard, not far from the *Great Eastern*'s slipway at Millbank, while the refitted *Great Britain* was working the England-

Scenes of celebration at the opening of the Royal Albert Bridge, Saltash, in May 1859. Brunel was too ill to attend the ceremony and shortly afterwards he was taken across the bridge on an open carriage drawn by a broad gauge locomotive. *Below*, the *Iron Duke* class of locomotives was introduced in the late 1840s. This replica of the eponymous *Iron Duke* is displayed at the Gloucestershire Worcestershire Railway in Toddington. Its massive 8 foot driving wheels give some idea of the size of these huge locomotives which could attain speeds of 80 mph on express services.

Punch's reverential portrayal of the ghost of Brunel
and the passing of the broad gauge in 1892.

Australia run for new owners. As for the *Great Eastern*; in truth this monstrous vessel had become an enormous white elephant before it was ever completed. Then, almost as a final straw, on 8 September 1859, Brunel received word that one of the *Great Eastern*'s feed-water heaters surrounding a forward funnel had exploded during trial runs. Several stokers were badly scalded in the accident, five of them fatally. It is said that Brunel never recovered from this dreadful news and he died one week later.

Triumph and disaster are often flip sides of the same coin. Despite the litany of lost works, this book is not meant as an attack upon Brunel's reputation or his achievements. Instead it is intended to fill in some of the gaps, looking beyond the familiar signature pieces in order to explore and celebrate the full breadth of his talents. There are various reasons why the works covered in this book were lost. Some projects were either never started or never completed, some works were not expected to last – in particular the wooden structures such as the Cornish viaducts and assorted station buildings – and a few simply failed to deliver. Of the remainder, many have been deliberately destroyed, including the water towers at Sydenham, while others have been swept away by engineering advances or the needs and demands of an ever-changing world.

In many respects it is shocking to discover just how much of Isambard Kingdom Brunel's work has been lost in the 150 years or so since his death. Even so, I think that Brunel, ever the pragmatist, would have been more amazed to see that so much has survived.

The moment of deluge as the Thames breaks in through Marc Brunel's tunelling shield. *Below*, contemporary illustrations of the newly opened Thames Tunnel invariably depict a pleasant scene with smartly-dressed pedestrians marvelling at this 'eighth wonder of the world'. But where is the horse-drawn traffic that the tunnel was meant to serve?

Like Father Like Son

Just as with George and Robert Stephenson, the Brunels came as a father and son set. Physically they were very similar, small men barely more than 5 feet tall, but their stature as engineers made them giants among their peers.

A refugee from the French revolution Marc Isambard Brunel had already established his reputation working as a surveyor and engineer in New York before he came to England in March 1799. His purpose in returning to Europe was twofold. Firstly, he was in love with a young English woman named Sophia Kingdom whom he had met in France. They were married later that year. Secondly, he was full of innovative ideas and in particular wanted to approach the British Admiralty with a proposal for the mass-production of ship's pulley blocks. A typical Royal Navy ship of the time required 922 blocks which, up until then, were made by hand; an expensive and time consuming business. Together with the mechanic Henry Maudsley, Marc Brunel built machines to manufacture these blocks more efficiently and in 1802 the Brunels moved to Portsmouth where Marc oversaw the installation of the first machines in the Naval Dockyard, where they remained in production well into the twentieth century.

It was in Portsmouth that the Brunel's third child and only son, Isambard Kingdom Brunel, was born on 9 April 1806. In Marc's mind there was no question that the boy would be an engineer like himself and, indeed, the young Isambard showed great aptitude which was carefully nurtured by his father. When he was still an infant the family moved to London to pursue Marc's business interests, including the mechanisation of sawmills and making boots for the British army which was at war with the French. Unfortunately, his sawmill at Battersea was destroyed by fire in 1814 and the demand for boots dried up after Napoleon's defeat at Waterloo the following year. Despite a period in the debtors' prison in Southwark, Marc bounced back and went on to work on a broad range of engineering projects, including designs for a cannon-boring mill plus two suspension bridges commissioned by the French government with one built on the Ile de Bourbon, off Mauritius, and the other at St Suzanne. However, it is the projects in which Isambard worked with his father that fall within the remit of this book, in particular the Thames Tunnel and the Gaz Engine. Both were magnificent failures.

In 1825 Marc began work on driving a tunnel under the Thames to provide access to the docks on the south side of the river as an

A nineteenth century depiction of the Thames Tunnel revealing how close its roof came to the river bed. The access shafts, for pedestrians only, are shown on either side, but the illustrator shows the tunnel as running dead level when it actually dips in the middle.

alternative to the overcrowded bridges. There had been previous unsuccessful attempts but his breakthrough idea was the 'Great Shield'. This consisted of twelve vertical iron frames, each 3 feet (0.9 metres) wide and divided into three cells, one above the other, with a miner working in each one. At the front of the shield was a series of horizontal boards which could be removed one at a time to expose only a small area to be excavated, thus reducing the risk of a cave-in. The boards were then moved forward and, once all the boards had been repositioned, the frame could be shunted forward by jacks pressing against the brickwork filling the gap behind the shield.

The first task was the construction of an access shaft at Rotherhithe on the north side of the river. Marc achieved this by building a huge 3 feet (0.9 metre) thick brick cylinder, 50 feet (15.2 metres) in diameter and 42 feet (12.8 metres) high, which was braced with vertical metal rods and had iron rims top and bottom. Acting like a massive pastry cutter this sank into the ground under its own weight as the ground was excavated beneath it. Once the shaft was completed the shield was put in place and the work on boring the tunnel itself commenced. Conditions for the miners, mainly recruited from the Cornish tin mines and the Durham coal fields, were appalling with the filthy river water constantly seeping in through the shield. The men wretched in the foul air and many were struck down by 'tunnel sickness', a condition which sometimes caused blindness. Inevitably, the terrible conditions took their toll and by April 1826 Marc's resident engineer, William Armstrong, had resigned. Marc was experiencing poor health himself by this time and there was only one candidate to take over the job.

IKB became resident engineer on Europe's most demanding engineering project when he was barely twenty years of age. He immediately threw himself into the task, spending more than twenty hours a day in the tunnel, with only cat-naps to keep him going,

often for days on end. By the beginning of 1827 progress was good and they had driven the tunnel 540 feet (165 metres), almost halfway to Wapping. But both Brunel's were increasingly concerned about the risk of breakthrough as instead of London clay the miners were encountering increasing amounts of gravel, often littered with river debris including clay pipes. Marc estimated that the shield was holding up to 600 tons of water at bay and his concerns were exasperated by the presence of the shilling-sightseers being admitted to the works by the company directors. Thankfully, there were no members of the public present when the river broke through on the evening of 18 May 1827. IKB was working a little way back from the shield when he looked up to see the panic-stricken workmen running for their lives. The surge of water extinguished the lights and in the blackness and deafening din he did his best to get everyone to the shaft as quickly as possible, but as they neared the top some of the steps were swept away. Suddenly a cry for help was heard and tying a rope around his waist he slid down one of the iron ties of the shaft to rescue an old engineman named Tillett.

Luckily, there were no casualties and IKB seems to have relished the excitement of his adrenaline-fuelled escape. He borrowed a diving bell from the West India Dock Company and descended beneath the cold waters to investigate the damage. By the end of the year the hole had been plugged with 150 tons of clay-packed bags and the workings were cleared in time for a celebratory Christmas banquet held in the tunnel itself. Work resumed, but on 12 January 1828 the tunnel flooded again. The torrent was so violent that the rush of water carried Isambard and his companions up to the lip of the shaft where he was plucked to safety by the others. He had sustained serious injuries to his leg and internally, but even so he called for the diving bell once more and directed operations from a mattress on the barge. The damage to the tunnel was far greater this time and with funds dwindling the directors ordered it to be bricked up. Isambard was sent to Brighton to recuperate and afterwards went on to Bristol to follow his own career. It was in an uncharacteristically despondent mood that he recorded in his journal:

You can still see the Thames Tunnel and even travel through under the river on the East London Line of the Underground. The twin horseshoe entrance is visible from the narrow platforms of Wapping Station, and the booking hall on the Wapping side incorporates the top of the access shaft. Take the stairs if you want to see the main part of the original brick-built shaft.

On the Rotherhithe side the pumping house, where steam engine operated to keep the tunnel dry, is now part of the Brunel Museum. The top of the Rotherhithe shaft is on the right of the picture. More recently, the base of this shaft, which had been retained as a ventilation shaft to the railway line below, has been closed off with a new floor to create further space for the museum's activities. (*The Brunel Museum*)

Tunnel is now, I think, dead... This is the first time I have felt able to cry at least for these ten years. Some further attempts may be made – but – it will never be finished now in my father's lifetime I fear.

He never worked on the tunnel again, but his predictions proved to be wrong. The Thames Tunnel was finally completed and officially opened to the public on 25 March 1843, only a few months before the launch of the *Great Britain*.

Marc received a knighthood from Queen Victoria in recognition of his endeavours, but the tunnel was never put to its intended use. 'The trouble was', as Robert Hulse, the director of the Brunel Museum at Rotherhithe, so eloquently puts it, 'they couldn't persuade the horses to go down the stairs!' Marc had intended to create two additional shafts, four times wider than the access shafts, with ramps to take the horses and wagons down to the tunnel level. Without these it was only fit for pedestrians paying a penny a head. When that novelty began to wear thin it became a sub-aquatic marketplace and degenerated into the haunt of thieves and prostitutes. Salvation of a sort came in 1865 when it was sold to the East London Railway to take the first underground line under the Thames.

Throughout his son's career and until his death in December 1849, Marc Brunel continued to advise and guide IKB on various engineering matters. There was, however, another notable joint project which has earned a place in this account of lost works. This was the 'Gaz Engine', an on-going series of experiments commenced by Marc and carried on by Isambard over a period of ten years. Inspired by Michael Faraday's lectures on the properties of gasses, the Brunels attempted to harness the expansive power of carbonic gas moving between a liquefied and a gaseous state. In effect they were trying to create a super-pressure form of steam engine. Marc devised an engine in which a double-acting piston moved back and forth within a cylinder. Linked to this were two pairs of pressure vessels and alternately one would be heated up causing the gas to expand as it returned to a gaseous state, while the other was cooled causing the gas to liquefy and condense. This would provide a pressure differential to drive the piston up and down to turn an axle and through gears drive an engine.

The theory was fine, but in practice they were dealing with gas pressures up to 1,500 lbs per square inch, roughly thirty times that encountered by conventional steam engines and well beyond the safe limits of the material available. It is nothing short of a miracle that they didn't blow themselves up in the process. Following the closure of the tunnel IKB erected the Gaz Engine apparatus in the abandoned works at Rotherhithe in order to continue with its development, but in the end even he had to admit defeat.

> Gaz – After a number of experiments I fear we must come to the conclusion that (with carbonic acid at least) no sufficient advantage on the score of economy of fuel can be obtained. All the time and expense, both *enormous*, devoted to this thing for nearly ten years are therefore *wasted* ... It must therefore die and with it all my fine hopes – crash – gone – well, well, it can't be helped.

As the historian Kenneth Clarke once observed, despite such setbacks, Isambard Kingdom Brunel 'remained all his life in love with the impossible'. Thank goodness he did.

Commemorating the opening of the Thames Tunnel in 1843, an illustration from *The Illustrated London News* with one of the miners and an oversized medal featuring Marc Brunel.

The Hungerford Bridge was commissioned to provide a pedestrian crossing from the south bank of the Thames to the Hungerford Market, *shown above*, situated on the northern side of the river on the site of the present Charing Cross Station. Work began on the bridge in 1841 and it was completed in 1845. With a central span of 676 feet, its twin towers, in an Itialianate style, rested on wide brick-built piers which provided access to the river boats. (*LoC*)

Hungerford Bridge

It has often been stated that while IKB rose to the challenge of a great project, whether a 'magnificent disappointment' or not, he often struggled to summon much enthusiasm for the lesser ones, the projects which either didn't allow him to exercise his engineering genius or accrue sufficient kudos. This was never more the case than with the Hungerford Suspension Bridge which he regarded as the poor relation to his 'first love', the bridge at Clifton. Ironically as it turned out, and entirely unknown to Brunel, the fates of these two suspension bridges were to be irrevocably linked.

By the time he was thirty years old Brunel was amassing an impressive array of engineering commissions including the design of a footbridge crossing the Thames connecting Lambeth, on the south bank, with the new Hungerford Market to be built on the northern side of the river. Any other engineer might have seized upon this as an opportunity to create a showpiece of engineering excellence right in the heart of the capital, albeit over a river which was little more than an open sewer and had become notorious as the 'Great Stink'. Brunel, however, already has his mind on bigger things and he dismissed the bridge out of hand, as illustrated by his journal entry for 26 December 1835:

> Suspension bridge across the Thames (Hungerford foot-bridge) I have condescended to be engineer of this but I shan't give myself much trouble about it. If it is done it will add to my stock of irons.

The commission for the Hungerford bridge had come via Brunel's brother-in-law, Sir Benjamin Hawes; a convenient example of nepotism or, as we prefer to call it nowadays, networking. Not surprisingly, in designing this suspension bridge Brunel drew heavily upon his experience gained with the designs for the Clifton Bridge. The main difference between the two was that the Hungerford Bridge was only intended for pedestrians and, consequently, the deck was only 14 feet (4.3 metres) wide. Brunel's proposal to widen it to accommodate carriages was not taken up. The bridge had a central span of 676 feet (206 metres), compared with Clifton's slightly bigger 702 feet (214 metres), with two long side spans of 343 feet (104.5 metres) each. This gave the bridge an overall length of 1,388 feet (423 metres) which is actually 36 feet (11 metres) more than Clifton. The two piers of the Hungerford Bridge were of hollow

Above and left: Two contrasting views of Hungerford Bridge: On the left is Fox Talbot's early photograph of the bridge in its prime, while the main picture, published in *The Illustrated London News* in February 1863, shows it during its conversion into a railway bridge. The south tower is being dismantled, the main deck with additional support columns is in place, and the vertical supports from the suspension chains have been severed.

Right: Beauty in the eye of the beholder, and while many Londoner's considered the Charing Cross Bridge to be hideously ugly, the painter Claude Monet found sufficient inspiration for a series of oil paintings produced during his visits to London between 1899 and 1903. Note the position of the main pier and also the Houses of Parliament in the background which corresponds with the *ILN* illustration, above.

Looking southwards along the new railway bridge, as seen from the end of the Charing Cross platforms in 1864. The 'signal station', running between the curved brickwork added to the top of the Brunellian piers, is also visible in the 1950s photograph of Charing Cross Station taken from the top of the Shell Building. *Compare this view with the one on page 25.*

construction in red brick dressed with stone and their wide footings served as landing piers for the river steamers with internal staircases leading up to the bridge level. On each pier stood a sturdy tower, in the Italianate style sometimes favoured by Brunel, and these supported the wrought iron chains that held the deck.

Work on the Hungerford Bridge had begun in 1841 and it opened on 1 May 1845. Despite the stink of the river it proved to be very popular with Londoners, especially following the opening of Waterloo Station on the south side of the river in 1848. It has been suggested that the opening of the bridge had provided some consolation to Brunel to make up for the abandonment of his beloved Clifton Bridge, but in truth Brunel had kept to his word and given the minimum of trouble to Hungerford's design. The result was functional, but lacked any great flair and, inevitably, the setting could never match the drama of the Avon Gorge at Bristol. This may account for the fact that the Hungerford Bridge was so easily forgotten; that and the extraordinary brevity of its existence.

The growing demand for railways into central London saw Parliament grant permission for the South Eastern Railway company to build a new station, Charing Cross, on the site of the Hungerford Market which was closed for business in 1859, the year of IKB's death. Hungerford Bridge was also sold to the railway company and the deck, towers and chains were removed, leaving only the original wide piers to be incorporated within a brutishly functional wrought iron girder bridge. This was designed by the South Eastern's engineer John Hawkshaw (later Sir John) and featured nine lattice girder spans resting on the piers plus additional pairs of iron columns.

The new Charing Cross Railway Bridge attracted a great deal of criticism initially and it was described as 'squat, ugly' and 'aesthetically notorious'. Because the railway company was required to maintain a pedestrian crossing, narrow walkways were cantilevered out from either side of the bridge and these continued as paid crossings until the toll was abolished in 1878. Later on, the upstream walkway was incorporated within the railway crossing, leaving only the eastern one for pedestrians and this earned itself a reputation as the haunt of muggers during the twentieth century. Clearly something better was needed and in the 1990s a competition was held to design new footbridges to run either side of the rail bridge. The winning design, by architect Alex Lifschutz, features slender suspension masts linked by steel rods to the walkways. They were completed in 2002 and although they are officially known as the Golden Jubilee Footbridges, in honour of the Queen's Golden Jubilee, they are more often referred to as the Millennium Bridges. However I am pleased to say that there is an increasing tendency to use the old Hungerford name.

It is estimated that seven million people use the twin footbridges to cross over the Thames every year, passing just feet away from what remains of Brunel's forgotten suspension bridge. Beneath the deck the brick and masonry piers are clear to see, although the curved red-brick and stone embellishments on the top of the piers are later additions. Curiously, the northern or Middlesex pier appears to be much closer to the shore than the other one, and this is because the northern river bank was extended with the construction of the Victoria Embankment in the 1860s to accommodate a new sewage system, water pipes and the tunnels for the Metropolitan District underground railway.

There is one last chapter to the Hungerford Bridge story which provides a direct link between Brunel's two suspension bridges. When Hungerford Bridge was being dismantled, John Hawkshaw also happened to be on the committee of engineers working to complete the Clifton Suspension Bridge as a memorial to Brunel. The wrought iron chains which had been originally made for Clifton, but never used there, had already been recycled by Brunel himself for the Royal Albert Bridge at Saltash. So Hawkshaw purchased the second-hand set of Hungerford chains for £5,000 and took them to Bristol where, augmented by an additional set to carry the greater load, they are now draped across the Avon Gorge. *See 'Clifton Suspension Bridge'.*

South Bank side of the bridge, photographed in the 1950s.

Charing Cross Station was given a new look with the addition of the Embankment Place development in the 1990s, as seen here from the London Eye. The railway bridge is still clearly visible, and you can see how the shoreline on the northern side moved closer to the original Brunel brick pier with the construction of the Victoria embankment in the 1860s. The railway bridge is flanked by the two Golden Jubilee Bridges, the first of which opened in 2002.

The Broad Gauge

At its height the broad gauge network stretched from London to Cornwall, and from the south coast to the Midlands and across through southern Wales. In terms of scale this must make it Brunel's biggest and costliest failure. Yet it is hard to condemn a system that was actually superior to the one it was eventually replaced by. This wasn't a case of a good idea giving way to a better one, it was all about who had the greater weight of numbers on their side in terms of territory covered.

When Brunel was appointed as engineer to the Great Western Railway in March 1833, he determined that it was going to be the finest railway in Britain. The railways were so new that there were no ground rules laid down by government or any other body regarding engineering details such as the choice of gauge, the distance between the rails. In the north of England George Stephenson had chosen the 4 feet 8.5 inch (1.4 metres) gauge for the Stockton & Darlington Railway simply because it was already being used in the collieries which it served, and naturally enough his son, Robert Stephenson, followed suit. That all seemed a long way from the Bristol to London railway that Brunel was designing. He dismissed the northern 'coal wagon' gauge in favour of a wider one at 7 feet and a quarter inch (2.1 metres) as he felt this would be 'more commensurate with the mass and velocity to be attained'. One of his early proposals had been that the railway carriages would be contained within the wheels rather than sitting above them, although this was never put into practice.

It has been suggested that Brunel chose a different gauge for the sake of being different from his rivals, but whether or not this is true the 'broad gauge', as it became known, was ratified by the GWR Board in 1835. The GWR would pioneer its own way and it was going to be magnificent. Brunel also had his own method of supporting the rails using longitudinal timber sleepers running the whole of their length, joined by cross sleepers at 15 feet (4.6 metre) intervals and anchored by beech piles driven into the ground. Once trains actually started running on these tracks it was discovered that the rigidity of the piling resulted in an extremely bumpy ride as the sections between them flexed up and down. Consequently, the piles were pulled up or driven down clear of the track.

Opposite: Replica of the *Iron Duke* broad gauge locomotive at the Gloucestershire Worcestershire Railway in Toddington.

When broad gauge and standard gauge met head-on pandemonium ensued, as shown in this chaotic scene at Gloucester. Brunel's solution was the transfer shed with broad gauge track on one side and standard on the other. Passengers, their luggage and all goods, had to transfer across from one train to the other to complete their journey. This preserved transfer shed is at the GWR Society's Didcot Railway Centre.

Over the next ten years, from 1835 to 1845, the various railway networks spread their spidery fingers across the map of Britain. The GWR extended westward beyond Bristol and the new lines connecting to it, with Brunel as their engineer, were also built to the broad gauge. It is sometimes overlooked that Brunel was not alone in adopting a 'non-standard' gauge. About a hundred miles of the London & North Western Railway had been constructed to a gauge of 5 feet (1.5 metres), some Scottish lines had been laid with 4 feet 6 inch (1.3 metre) and 5 feet 6 inch (1.7 metre) gauges, and the Surrey Iron Railway was of 4 feet (1.2 metre) gauge. Opinions varied on the merits of the various gauges, both in terms of performance and the desirability of standardisation but matters only came to a head when different gauges met head on. This happened for the first time at Gloucester where the Midland Railway met an extension of the GWR. Brunel's solution was the transit shed which had broad gauge track entering on one side of a central platform and the 'standard gauge', as it was already being referred to, on the other. All passengers, their luggage and goods had to be laboriously transferred from one train to the other. This resulted in pandemonium, as described by *The Illustrated London News*:

> It was found at Gloucester that to trans-ship the contents of one wagon full of miscellaneous merchandise to another, from one Gauge to another, takes about an hour; with all the force of porters you can put to work on it ... In the hurry the bricks are miscounted, the slates chipped at the edges, the cheeses cracked, the ripe fruit and vegetables crushed and spoiled; the chairs, furniture, oil cakes, cast-iron pots, grates and ovens all more or less broken... Whereas, if there had not been any interruption of gauge, the whole train would in all probability have been at its destination long before the transfer of the last article, and without any damage or delay.

The government called for a Royal Commission to look into the matter. Forty-four witnesses gave evidence on behalf of the standard gauge, while the broad gauge was defended by only four, including Brunel and Gooch. Trials between the two factions were conducted with trains travelling between Didcot and London, and the broad gauge came out on top. But with considerably more miles of standard gauge than broad, the Commissioners recommended that for the sake of uniformity the 4 feet 8.5 inches (1.42 metres) should be declared the gauge for all public railways. In 1846, the Gauge Act dictated that all new lines had to conform to the standard gauge, unless they were extensions to the existing broad gauge network. Although the use of mixed gauge lines kept the broad gauge trains working for another fifty years, but this had many drawbacks not least of which was the cost, and gradually the broad gauge lines were converted to standard. In 1892 the remaining 415 miles (668 km) of broad gauge, between London and Penzance, were converted in an extraordinary show of organisation lasting thirty-one hours over the weekend of 21–22 May.

For many the broad gauge represents a lost opportunity. A *Punch* cartoon portrayed the ghost of Brunel passing among the navvies as they carried out their work. Entitled 'The Burial of the Broad Gauge' it bade farewell with the words, 'Good-bye, poor old Broad Gauge, God bless you.'

Brunel and locomotives

It is a curious fact that Brunel's genius did not extend to the design of locomotives. When it came to placing the orders for the first locomotives he imposed a set of specifications upon the builders that could only result in an inferior performance. A speed of 30 mph (48 km/h) was to be considered the 'standard velocity', and at that rate

This section of mixed gauge track at Didcot Railway Centre illustrates the difference between the outer pair of broad gauge rails, and the middle and right-hand rails of the so-called standard gauge.

First and last: *North Star* was produced by Robert Stephenson's Newcastle company for an American customer originally. It was converted to Brunel's broad gauge for the GWR and on 31 May 1837 it worked the inaugural train. This replica was built to mark the GWR's centenary in 1935, but only the driving wheels are from the original. *Bulkeley*, an *Iron Duke* or *Rover* class 4-2-2 named after Captain Bulkeley, one of the company directors, represents the final generation of broad gauge locomotives. On 20 May 1892, it hauled the last broad gauge train out of Paddington.

The end of the broad gauge as the *Bulkeley*, travelling westwards out of Paddington, pauses for a farewell photograph in Sonning Cutting, 20 May 1892. That weekend the remaining broad gauge mainline track was converted to standard gauge in a mammoth operation involving thousands of men. The unwanted broad gauge locomotives – they couldn't be converted – were lined up in this graveyard at the Swindon Works to await scrapping. (*STEAM Museum of the GWR*)

the piston speed should not exceed 280 feet per minute, whereas 500 feet per minute was not at all unusual on the standard gauge locomotives already in service. Consequently, the first examples built to his requirements and delivered in early 1838, featured excessively large driving wheels powered by undersized boilers. They proved to be underpowered and unreliable, and in the ensuing furore Brunel's reputation and his job were on the line. L. T. C. Rolt, Brunel's most celebrated biographer, didn't pull any punches when he summed up the situation:

> For the motive power troubles which beset the Great Western in its earliest days Brunel was far more culpable; indeed it is safe to say that these first locomotives represent the most inexplicable blunder in his whole engineering career.

Fortunately salvation came in two forms. Firstly, the GWR got its hands on *North Star*, a 2-2-2 locomotive built by the Robert Stephenson Company in Newcastle. *North Star* had been intended for a railroad company in New Orleans which went out of business before the locomotive could be shipped out to them. It was converted from the American 5 feet 6 inch (1.7 metre) gauge to Brunel's broad gauge and it was *North Star* that hauled the GWR's inaugural train on 31 May 1838. Secondly, a brilliant young engineer named Daniel Gooch joined the company as Locomotive Superintendent. Gooch's superior skills as a mechanical engineer must have rankled with Brunel at first, but over the years they became close friends.

Brunel's great broad gauge experiment cost the GWR dearly. Following the gauge conversion in 1892 the engineering works at Swindon became a graveyard for the unwanted broad gauge locomotives waiting to be scrapped. The *North Star* and *Iron Duke* escaped this fate for a while, but in 1906 they too were scrapped because they occupied too much space in the workshops. Only *North Star*'s 7 feet (2.1 metre) driving wheels were kept and these formed the basis of a replica built to mark the GWR's centenary in 1935; it now resides in the GWR Steam Museum in Swindon. Other replicas have been built including an *Iron Duke*, which is, at the time of writing, on display at the Gloucestershire Warwickshire Railway at Toddington. There is also the *Fire Fly*, a working replica based at the GWR Society's centre in Didcot. The only genuine broad gauge loco to survive is a small vertical engine known as *Tiny* which is at Buckfastleigh in Devon. However, the greatest and most enduring monument to the broad gauge is Brunel's magnificent railway line which runs from London to Bristol.

The First Paddington

The GWR's London terminus at Paddington is rightly regarded as one of Brunel's architectural triumphs, a lofty cathedral-like space enclosed by three elegant naves of iron and glass. Unknown to most travellers, however, when it opened in 1854 this station was known as Paddington 'New' Station for the simple reason that it had replaced an earlier one.

For the fledgling railway companies of the 1830s, finding a suitable location for a major station on the edge of London's rapidly spreading built-up area was not a simple matter. Brunel considered several alternatives, including Paddington, but because of difficulties in obtaining land to build the line into the capital, the company's directors decided to approach their rivals at the London & Birmingham Railway (L&BR) with a view of sharing facilities with a junction at Wormwood Scrubs and a joint line going into Euston. Brunel met with Robert Stephenson, the L&BR's engineer, to discuss how the station could be split between the two companies, but in the end the GWR's directors pulled out as they were concerned that the L&BR would have too much influence on their operations and also because of the question of competing gauges.

With Paddington back on the agenda Brunel set to work drawing up plans for a station building fronted by an imposing classical stonework facade and featuring a central archway for horse-drawn carriages leading through to a roadway between the platforms. It had been common practice for wealthy passengers to travel in their carriages loaded on to the railway wagons. By the time Parliament had given its approval for a terminus at Paddington, the cost of constructing the GWR was already way over budget and Brunel's grand plans were swept aside. Work began instead on a temporary station on a site just to the north-west of the present one, located within a shallow bowl tucked beneath the Grand Junction Canal and known as the Paddington Basin. Most of the land belonged to the Bishop of London, hence the name of the Bishop's Road Bridge which the GWR was required to build across the site. This brickwork bridge became a central feature of the original station and some of its arches housed the booking hall, waiting rooms, parcel office, cloakrooms and the entrance and exit to the platforms.

Opposite: Detail from William Powell Frith's celebrated painting of Paddington New Station, c. 1862. The original Paddington station had been little more than a cluster of wooden huts.

Looking northwards towards the Bishops Road Bridge with nine of its arches serving as the entrances to the original and temporary Paddington Station. This opened in June 1838 and in J. C. Bourne's lithograph, above, the scene has been tidied up somewhat, leaving out the goods shed and offices which would have been on the left. In the second view, looking westwards towards the Prince of Wales Hotel, we can see the cottage-like goods office and this time the station approach is full of the hustle and bustle you might expect at a busy railway terminus.

The station itself consisted of a number of wooden platforms protected from the elements by simple timber-built roofs supported on slender iron columns. There were five lines of track with the three platforms for departures separated from the arrivals platforms by a roadway for carriages. At the end of the platforms was a series of wagon turntables and beyond these were the main carriage shed and an engine house. The latter was designed by the GWR's young locomotive superintendent, Daniel Gooch, and featured an innovative polygonal 'roundhouse' with a central turntable – the first engine shed to be built with this layout. Seen from the end of today's platforms at Paddington, the main part of the original station was located just beyond the Bishop's Road Bridge, with only the goods shed and goods shed office on the nearer side accessed by the trains via the arches on the left-hand side. The timber-built goods shed was 330 feet (101 metres) long and 120 feet (37 metres) wide. Standing beside it was the goods office which has been described as looking like a converted cottage.

The station opened for business in June 1838, although initially the trains only ran as far as Maidenhead. The line was incrementally extended westwards as it was completed in stages. There are very few contemporary descriptions of the old station, although in 1843, when Prince Albert came to Paddington to catch a train to Bristol for the launch of the *Great Britain* steamship, *The Illustrated London News* published two engravings. We also have J. C. Bourne's lithograph of the Bishop's Road Bridge facade, and together these provide us with the most detailed visual record. Bourne confirmed the temporary nature of the station, stating;

> The present arrangements at Paddington are temporary only; the plot of ground already excavated to the south of the booking-office, and partially occupied by a large goods shed, has been purchased by the Company, and is understood to have been set aside for the purposes of a permanent station.

Unfortunately, the 1843 report in the *ILN* didn't provide any more details on the facilities:

> At the moment of the arrival of His Royal Highness at the terminus the scene was truly animated. The reader will recollect that although the railway company has not yet constructed an expensive terminus, they have a very large establishment here for the repair of engines and carriages, and every convenience for carrying on the traffic. As this is the longest independent line of

railway completed in this country, so all its appointments are in keeping with this superiority.

Clearly that last bit did not refer to the wooden station which was proving to be woefully inadequate to meet the demands of the increasingly busy railway. In 1850, the company directors gave the go-ahead for the construction of a new terminus on the plot of land south of the Bishop's Road Bridge. Not surprisingly, Brunel greeted the opportunity with enormous enthusiasm:

> I am going to design, in a great hurry, and I believe to build, a station after my own fancy; that is, with engineering roofs, etc, etc. It is at Paddington, in a cutting, and admitting of no exterior, all interior and all roofed in.

For Brunel the rebuild was a unique opportunity to rethink his approach to station design. Much had happened since he sketched out his first plans almost twenty years earlier and now he drew heavily upon Jospeh Paxton's work using iron and glass for the Great Conservatory at Chatsworth House and also the building for the Great Exhibition of 1851, which was christened by *Punch* as the 'Crystal Palace' – *see 'The Two Towers'*. The present Paddington station is unique among the London Termini as it has no grand facade, this role being fulfilled by the Great Western Royal Hotel on Praed Street, designed by Phillip Charles Hardwick in the French Second Empire Style.

The first Paddington Station was demolished around 1852. The land it occupied became part of the GWR Goods Station, and this in turn has been demolished to make way for the Sheldon Square complex of offices and apartments. The original brick-built Bishop's Road Bridge was replaced with a steel girder bridge before the First World War, to improve the line of the tracks leading into the present station. This girder bridge was replaced in turn by the present road bridge in 2005.

This engraving of the original Paddington, showing the covered walkway between the departures and arrivals platforms, was published by *The Illustrated London News* in 1843. Itsshows the simple wooden roofs supported on slender iron columns. In stark contrast the new Paddington, below, is worlds apart. This is a cathedral-like space defined by three lofty and lightweight transepts of wrought iron and glass.

Brunel's other bridge over the Avon in Bristol: The 1839 railway bridge coming into the city at Brislington, on the approach to Temple Meads, is a fine Grade I masonry structure with a wide central arch and two smaller flanking arches in a Gothic style. The only trouble is that you can't see it because of the later girder bridges on either side. This is also a very awkward site to access and the only view is from the hideous concrete bridge on Feeder Road.

Railway Works

With an estimated 1,200 miles (193 km) of railway line to his credit, it is no surprise that the greatest proportion of Brunel's lost works falls within the category of railway structures, in particular the functional, workaday buildings such as the numerous stations or goods and engine sheds. In many cases these were never intended to last forever as they were constructed of timber, just as with the first Paddington Station. But even Brunel failed to envisage the extent of the increase in rail traffic by the end of the nineteenth century and many of the original stations and other facilities were lost to the continual expansion of the railways, in particular on the main GWR line from London to Bristol.

Railway Stations

As it is beyond the scope of this book to catalogue all of the lost stations on the myriad of railway lines designed by Brunel or his team of assistants, this section will concentrate on the London to Bristol line. This had six principle stations; Paddington, Didcot, Swindon, Chippenham, Bath and Bristol Temple Meads, although two of these weren't opened until after the through trains had already commenced operation; Swindon in 1842 and Didcot in 1844. Apart from the obvious exception of Paddington, for the most part the principle stations have survived intact, although Bristol's terminus buildings, including the magnificent wooden roofed passenger shed and the train shed, ceased to be used for trains in 1965. The station at Bath had a similar wooden roof to Bristol's, complete with hammerhead-beams, but sadly this was removed in the 1890s. As can be seen from J. C. Bourne's lithograph of Bath Station, it originally accommodated four tracks of broad gauge. Now minus its roof and the middle tracks the station is left with an exceptionally wide space between the platforms.

With the benefit of over 150 years of hindsight the layout of a station with platforms on either side of the track for the up or down trains seems obvious. However, we should not forget that Brunel and his contemporaries were pioneering a whole new form of transportation. They were working with a blank sheet of paper and sometimes they got it wrong, as was the case at Reading and Slough where Brunel came up with an unusual layout for single-sided stations, although in fairness there are mitigating circumstance to explain this design. At the time of constructing the railway the two towns were both largely located to the southern side of the line. Furthermore, Brunel was under pressure from the GWR's London directors to curb his architectural excesses in to order to keep a lid on the escalating costs. A one-sided station consisted of two platforms set slightly apart on the same side of the track. In effect these were two stations, one for up trains and the other for the down trains. The result was an excessively complicated track layout – best seen on the diagram of Reading Station – and also a potentially dangerous one with four places where the lines crossed. Consequently, the one-sided stations were replaced in the 1890s.

The remainder of the stations on the main line were classified as minor stations and typical of these was the one at Pangbourne, consisting of a single storey building with an all-round canopy, a pitched roof and signature Brunellian skew chimneys in a vaguely Elizabethan style. Bourne's lithograph shows the station building, containing the booking office and waiting rooms, situated on one side of the tracks with a simple shelter on the other side. To save money the design of the minor stations was standardised and this example became known as the 'Pangbourne type'. All of the minor stations on the London to Bristol mainline were demolished as a result of the quadrupling of the tracks in the 1890s and the withdrawal of stopping trains during the twentieth century. Fortunately, there is one surviving example of a Pangbourne type station at Culham on the Didcot to Oxford branch line which escaped the increase in tracks. There is a similar station at Mortimer on the former Berks and Hampshire line, although this example has a wider low-pitch roof which extends to the edge of the awnings.

Note that there were no footbridges between the two platforms on these minor stations originally, and passengers were expected to descend on the platform steps in order to cross the railway tracks. This may have been acceptable in the early days with fewer trains, but footbridges were added later on for obvious reasons.

Other railway buildings

Apart from the stations, the railway companies also needed a wide range of other buildings, from offices to engine and goods sheds, from major construction and repair facilities to water towers, coaling yards and a host of minor track-side buildings. The largest GWR facility was at Swindon, a green field site selected by Gooch as it was roughly the halfway point between London and Bristol, as well as being the dividing point between the low gradients from the London end of the line and the more challenging terrain going westwards to Bristol. In 1841, construction began on a group of buildings including an engine shed, plus engine repair and erecting shops. The biggest of these, the Engine Shed, was 490 feet (149 metres) long and 72 feet (22 metres) wide; big enough to accommodate up to forty-eight engines and

As revealed by J. C. Bourne, Bath Station originally had an all-over wooden roof much in character with the one at Temple Meads. The interior view shows the roof, already dirty with soot, and the side columns. There are four tracks of broad gauge, nowadays reduced to two of standard gauge, and note the railway wagons, in particular the one on the right which appears to be for luggage only. The surprisingly rural exterior scene shows the station on the right with the St James Bridge coming in from the left, a view which is now completely obscured by buildings.

Reading Station, and later the one at Slough, were oddities in that they were both one-sided with the 'down' and 'up' platforms located on the southern side of the line. Brunel's experiment in station layout may have been prompted by pressure from the company directors to keep costs down, but as can be seen from the complicated track layout, *below*, it was a recipe for disaster with the tracks crossing over each other in several places. Both stations were later replaced with more conventional layouts.

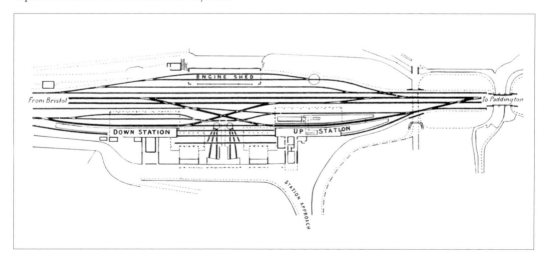

tenders. At right angles to this was the Locomotive Repairing Shop, a building immortalised in J. C. Bourne's series of GWR lithographs. 290 feet (88 metres) long and twice the width of the Engine Shed, the Repairing Shop featured a central transverser, to move the locomotives, with side bays on either side. The building was constructed with masonry walls and a triple roof of timber with a double-rise centre section braced with iron ties and supported on timber columns, which were later replaced with iron ones. Swindon Works grew very quickly, doubling in size in the first five years with a corresponding increase in the workforce from 423 to 1,800. This expansion continued into the twentieth century and the Engine Shed and Locomotive Repairing Shop were demolished in 1929 for make way for newer workshops.

Brunel would have been the first to recognise the need to improve the railway's facilities. In another of Bourne's illustrations we get a glimpse of the old goods shed at Bristol Temple Meads, built at right-angles to the main station building, and located between it and the Floating Harbour for the transfer of goods to and from barges. This building was open on one side and featured a fine timber roof tied with

For the smaller or secondary stations on the GWR, Brunel produced a standardised design known as the 'Pangbourne' type. This featured a single storey building with steeply pitched roof and skewed chimneys, plus a smaller shelter on the other platform. All mainline secondary stations were swept away by the enlargement of the railway, although examples have survived on branch lines.

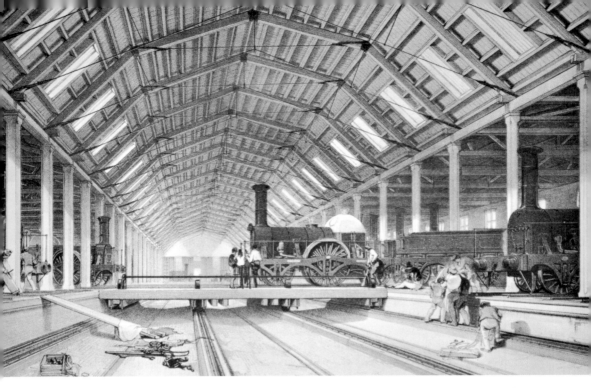

Swindon's Locomotive Repairing Shop was a 290ft long building with masonry walls and a triple roof of timber. It featured a central transverser to move locomotives to and from the side bays. Inevitably it fell victim to the rapid expansion of the Works and in 1929 both the Engine Shed and Locomotive Repairing Shop were demolished. In this 1927 photograph, *below*, a King Class boiler is being lowered on to the driving wheels in the Erecting Shop. (*STEAM Museum of the GWR*)

iron rods. As it was slightly lower than the main line into the station the railway wagons had to be raised or lowered by hoists. In the 1870s, the goods shed was demolished as part of the rebuilding and expansion of the Temple Meads complex, although, thankfully, the Passenger Shed, Train Shed and offices did survive – *see 'Lost and Found'*.

Unusually for a railway engineer, Brunel, with a keen eye for detail and a love of architecture, was closely involved in every aspect of the design of the stations as well as many of the more functional buildings, although on the later branch lines these were largely in the hands of his assistants. The station at Frome, on the former Wilts, Somerset & Weymouth Railway, provides us with a remarkably intact example of a timber roofed broad gauge station which, it is known, was designed by his assistant J. B. Hannaford. Likewise, the station and goods shed at Stroud, on the former Cheltenham & Great Western Union Railway from Swindon via Cirencester across to Gloucester, is probably also the work of one of Brunel's assistants.

These examples are a remainder not only of the complications in attributing particular works to Brunel, but also of the vulnerability of much that survives. The high profile buildings and bridges might be protected, but there is a plethora of unidentified smaller works that remain at threat. The old goods shed at Frome has disappeared in the last twenty years or so, for no apparent reason as the land it occupied has not been reused. Elsewhere there is better news, such as at Stroud, for example, where the goods shed is to be used as a community arts venue.

Railway bridges

Just as with the railway stations, many railway bridges have fallen victim to the increased levels of railway traffic and, in particular, the increased weight of the trains. This includes the loss of several high profile bridges, notably Chepstow and the wooden skew bridge at Bath, plus all the wooden viaducts in Cornwall, which are covered in other sections of this book. Perhaps what is remarkable is that so many of the bridges have survived at all despite the quadrupling of the track and subsequent increase in width of many bridges and viaducts, such as the Wharneford Viaduct spanning the Brent Valley, plus the trio of brick-built Thames bridges at Minehead, Basildon and Moulsford – each of them incredible structures in their own right. Elsewhere there are other interesting survivors, most notably the wrought iron bowstring bridge at Windsor.

In comparison to the other great bridge builders, Thomas Telford and Robert Stephenson, Brunel had a particular distrust for cast iron as a building material. This should not be confused with wrought

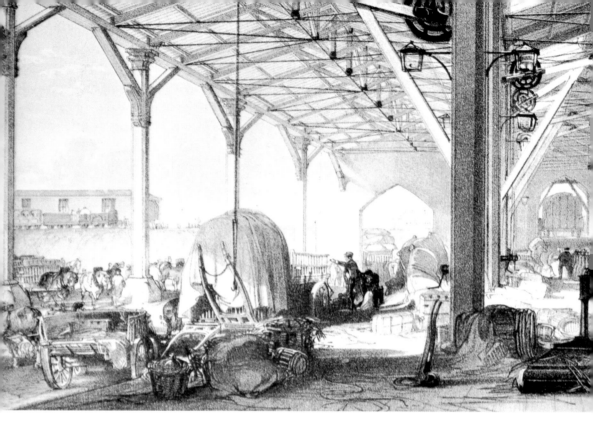

iron which is much stronger and less brittle than cast iron but didn't
become widely available until later in his bridge-building career. This
dislike of cast iron may stem from his experiences with the railway
bridge over the Uxbridge Road at Hanwell; an awkward site where the
line had to pass over the junction where two turnpike roads crossed
at a very sharp angle. It was too sharp for a brick bridge and instead
Brunel devised a bridge with iron beams supported on masonry
abutments at either side, and sixteen hefty columns underneath.
During the construction of the bridge in 1837 one of the girders broke,
and then another only a year after the railway had opened. In 1847, the
wooden decking caught fire and the ironwork was so badly damaged
that the entire bridge had to be rebuilt.

The Goods Shed at Bristol Temple Meads was at a right-angle and slightly lower than the main station. At the far end of the building a wagon can be seen being raised to the station's level. Below, the bridge across the Uxbridge Road coincided with an awkward junction and this led Brunel to put aside his distrust of cast iron. Unfortunately one of the girders broke and, later on, the wooden decking caught fire.

A Transatlantic Vision

Isambard Kingdom Brunel was an incurable visionary. Even before the first trains had run on the Great Western Railway his thoughts were already turning to an integrated international transport system. The genesis of this grand concept is shrouded in a haze of cigar smoke and its telling has become one of those apocryphal moments in the Brunel story. At a meeting of the directors of the GWR held in Blackfriars in 1835, one of them happened to comment on the inordinate length of the line. In response Brunel exhaled a curl of bluish smoke and suggested, 'Why not make it longer, and have a steamboat go from Bristol to New York and call it the *Great Western*?' And that is just what they did.

In January 1836, the Great Western Steamship Company was formed. It wasn't part of the GWR but the directors did include Brunel's colleagues and he was appointed as engineer, giving his services for free. In most respects the wooden-hulled *Great Western* followed conventional lines, although at 212 feet (65 metres) long and with a displacement of 2,300 tons she was considerably bigger than any other ship afloat. The Bristol ship-builder William Patterson was selected to build the ship at his yard near the Prince Street Bridge, and in June 1836 the keel was laid. Because of her length great emphasis was placed on longitudinal strength and extra stiffening was provided by iron diagonals and a row of iron bolts running the full length of the hull. The *Great Western* was driven by a pair of paddle wheels, assisted by sail as was conventional with the early steamships. However, there were many naysayers who thought that the notion of a transatlantic steamship was pure folly. Enter Dr Dionysius Lardner, an outspoken critic who targeted Brunel at every opportunity:

> As to the project of making the voyage directly from New York to Liverpool, it was perfectly chimerical, and they might as well talk of making the voyage from New York to the Moon ... 2,080 miles [3,347 km] is the longest run a steamer could encounter – at the end of that distance she would require a relay of coals.

Dr Lardner had failed to grasp a rudimentary factor in the design of large ships and he was convincingly rebuked by Brunel:

Opposite: The *Great Britain* flying the same flags as on the day of her launch in 1843.

In 1837 Brunel's first steamship, the *Great Western*, was launched from Patterson's yard in Bristol. The ship was intended to provide the final link in a transport system stretching via the railway from London to Bristol, and then across the Atlantic by steamship to New York. The Great Western Hotel, in St George's Street, designed by R. S. Pope in collaboration with Brunel, was built to accommodate the travellers. Carriages entered through the arches to a courtyard at the back. In the event, the *Great Western* never operated out of Bristol's docks and the city lost its transatlantic trade to Liverpool.

The resistance of vessels does not increase in direct proportion to their tonnage. This is easily explained; the tonnage increases as the cubes of their dimensions, while the resistance increases about as their squares; so that a vessel of double the tonnage of another, capable of containing an engine of twice the power, does not really meet with double the resistance. Speed therefore will be greater with the large vessel, or the proportionate power of the engine and consumption of fuel may be reduced.

The *Great Western* was launched on the morning of 19 July 1837 cheered on by a crowd 50,000 strong. With her large paddle wheels carried on the upper deck she sailed to the Thames for final fitting out. Meanwhile the Great Western Hotel was being built in Bristol, located in St George's Road tucked behind the present Council House, to accommodate the passengers who would travel down by train from Paddington to catch the steamship to New York. The first advertisements for transatlantic sailings appeared in March 1838 announcing that the 128 state rooms were all of one class with a fare of thirty-five guineas, with twenty 'good bed places' allocated for servants who travelled at half price. That same month the *Great Western* was put through her first steam trials, but not without incident. Lagging on the boilers caught fire and Brunel fell through a charred ladder into the boiler room. The ship returned to Bristol, but significantly not to the city docks as the entrance lock was too narrow now that her paddle wheels were in place, and alternative moorings were arranged at Pill further up the Avon.

In April 1838 the *Great Western* set off for New York on her maiden transatlantic voyage. She wasn't the first steamship to make the Atlantic crossing as she had been pipped to the post by the smaller *Sirius* which had sailed from Liverpool several days earlier. But it was the *Great Western* that established a regular transatlantic service, although with Bristol soon out of the picture. The Bristol Dock Company had failed to heed IKB's calls for improvements to the entrance locks and the city lost out to Liverpool.

Undeterred, the GWSC started work on a second vessel. Initially to be called *The City of New York*, this revolutionary ship eventually became the *Great Britain* when she was launched on 19 July 1843, six years to the day after the *Great Western*. With a hull of wrought iron and propelled by a screw propeller the *Great Britain* has been rightly described as the first modern ocean liner. At 322 feet (98 metres) long and with her six tall masts, named after the days of the week, she was a magnificent sight for the thousands of onlookers who flocked to witness her launch.

Brunel had a hand in making several improvements to Bristol Docks, including the South Entrance Lock which is on the right of this photograph. However despite assurances from the Docks Company the entrance from the river into the Cumberland basin was not improved in time for the *Great Britain*'s departure in 1844. His transatlantic ambitions were finally scuppered two years later when the *Great Britain* ran aground at Dundrum Bay on the Irish coast, *left*.

Strictly speaking the *Great Britain* cannot be categorised as one of IKB's lost works, only 'almost lost' – *see 'Lost and Found'* – but there were two incidents that impacted enormously upon the fate of the *Great Western* and on Brunel's transatlantic vision. After her launch the *Great Britain* remained in Bristol to be fitted out until she was ready to commence sea trials the following year, 1844. On the first attempt to get her through the Cumberland Basin and out to the river the hull became stuck as the docks company had not widened the locks as expected and also because she was riding slightly lower in the water than planned owing to the heavier steam engines which had been installed. On the second attempt Brunel had some of the dockside masonry temporarily removed and she just scrapped through on the high tide. Quite clearly she would not be coming back to Bristol, at least not during her active career.

The *Great Britain* made her first transatlantic run from Liverpool to New York in 1845, sailing with just fifty passengers on board as there were widespread concerns about the method of construction and the effect that the iron hull might have upon the compass. Brunel thought the latter problem had been solved by a system of correcting magnets devised by the Astronomer Royal. Passenger confidence gradually picked up and by the fifth voyage, in September 1846, she sailed with 180 passengers. Unfortunately, in the darkness and driving rain the ship ran aground at Dundrum Bay, on the Irish coast in County Down. Captain Hoskin claimed he had been confused by a newly commissioned lighthouse on the southern tip of the Isle of Man, but it is quite plausible that the compass was to blame. Thankfully there was little damage to the hull and Brunel devised protection for the vulnerable vessel until she could be re-floated the following summer. The *Great Britain* was saved, but the GWSC could not survive the financial repercussions of this incident and both ships had to be sold.

The *Great Western* was bought by the West India Mail Steam Packet Company in 1847 and was used on the West Indies route and also she served briefly as a troopship during the Crimean War. In 1857, the ship was broken up at Castles' Yard in Millbank, not far from where Brunel's final ship the *Great Eastern* was being built. It is said that the engineer went there to take a farewell look at her before she disappeared. It was just twenty years after her launch.

Brunel's docks

As a civil engineer Brunel was involved in dock improvement projects at a number of locations: At the entrance to the River Wear at Monkwearmouth, Sunderland; the Great Western Dock at Plymouth; Briton Ferry in Baglon Bay, South Wales; Brentford Docks on the

Thames; Neyland Docks, New Milford in Pembrokeshire; and at Bristol which Brunel had hoped would become the hub of his international transportation system.

Bristol's Floating Harbour had been created in 1804 by enclosing part of the river between the Neetham Dam at Temple Meads and Rownham Dam on the western side of the city. To accommodate ships entering and leaving the 'Float', as the dock was known, the Cumberland Basin was created to link the docks and the Avon. Unfortunately, the reduced water flow within the Float resulted in a problem with accumulated mud deposits and in 1832 the Docks Company asked Brunel to make several improvements. These involved installing sluice gates and creating an overflow or siphon running out to the river through the Underfall Yard. Later on Brunel designed a steam-powered drag boat to scrape mud out of the Cumberland Basin. His most visible contribution can be seen at the south entrance to the basin where he enlarged the lock and constructed an innovative tubular iron swing-bridge to carry the road traffic over the entrance – *see 'Chepstow's Tubular Bridge'*. Brunel also devised a floating pier to receive large ships at Portishead, near the mouth of the Avon, but this never came to fruition.

The *Great Britain* could so easily have joined the list of lost works. In the 1930s she was abandoned at Sparrow Cove in the Falklands and left to rot. If she hadn't been recovered and returned to Bristol in 1970 there is every chance that the sea water and the rust would have finished her off by now. (*CMcC*)

The Atmospheric Caper

One of Brunel's greatest skills was his ability to learn from the work of his contemporaries and to adopt it for his own purposes, as had been the case with the use of screw propulsion on the *Great Britain*. In September 1844, he travelled to Ireland with Daniel Gooch to witness a demonstration of the atmospheric system on the Dalkey Kingstown Railway. This single track, 1.75 miles (2.7 km) long, had been built by Charles Vignoles using a principle pioneered by the gas engineer Samuel Clegg and the Samuda brothers, Jacob and Joseph. The Clegg-Samuda system consisted of a cast-iron pipe, 15 inches (38 cm) in diameter, laid between conventional rails. The pipe had a slot running its length on the upper side, and inside the pipe was a piston which was connected via a metal arm through the slot to the wagons. The slot was closed by means of a leather flap which was greased to make an airtight joint at one edge and hinged at the other. Stationary pumping engines situated alongside the track pumped the air out of the pipe ahead of the train and the atmospheric pressure behind the piston pushed it along.

Brunel was immediately smitten by the notion of 'harnessing the atmosphere' as motive power. For him the advantages were obvious. The lighter atmospheric trains could cope with greater gradients than heavier steam locomotives and consequently, he argued, an atmospheric railway would be much cheaper to construct. He stubbornly ignored the opposing views held by Daniel Gooch and Robert Stephenson, who happened to be two of the country's most capable mechanical engineers. Gooch later wrote:

> I could not understand how Mr Brunel could be so misled. He had so much faith in his being able to improve it that he shut his eyes to the consequence of failure.

Stephenson also dismissed the atmospheric railway, calling it a 'great humbug' and 'a rope of wind'.

The South Devon Railway received Parliamentary assent in July 1844 and the initial suggestion that the atmospheric system might be considered for the line between Newton Abbot and Exeter had come from Clegg & Samuda. As engineer to the line Brunel added his enthusiastic endorsement.

> I have no hesitation in taking upon myself the full and entire responsibility for recommending the adoption of the atmospheric

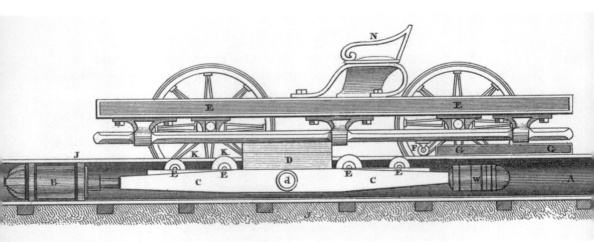

The saga of the atmospheric railway has a modern equivalent in the Sinclair C5 electric car. They both sounded wonderful in theory, but in practice they were engineering disasters. Atmospheric propulsion was pioneered by Samuel Clegg together with Jacob and Joseph Samuda, and these drawings illustrate the mechanics of the system. A piston travels along a metal tube and is driven by atmospheric pressure when the air in front of it is evacuated by a stationary steam engines. The piston is joined to the underside of the train by a connector passing through a slot in the top of the tube, and this slot is kept airtight with a leather flap.

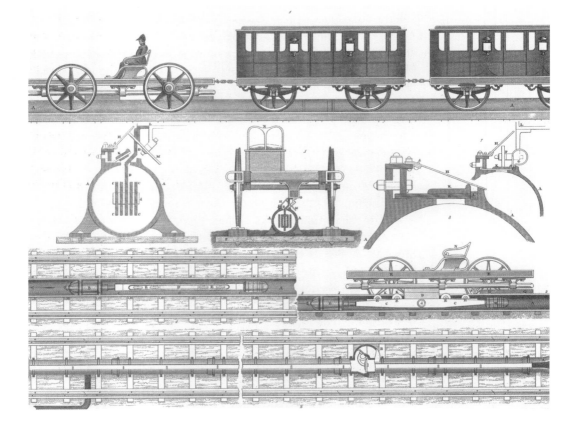

system on the South Devon Railway and recommending as a consequence that the line and works should be constructed for a single line only.

As usual his strength of purpose over-rode any objections from the company's directors, who had signed up for a conventionally hauled double-track line. By February 1847, the railway line between Exeter and Teignmouth had become operational, although it was initially worked by steam trains until August when the atmospheric system was completed. By January of the following year it was extended to Newton Abbot. Along the length of the 20 mile (32 km) track eight stationery steam engines were installed in engine houses, finished in an Italianate style and located at approximately 3 mile (4.8 km) intervals.

At its best the atmospheric system resulted in trains that could run at speeds of up to 70 mph (113 km/h), although in practice the average travelling speed was a more modest 40 mph (64 km/h). The passengers enjoyed the novelty of smoke and cinder free rail travel, but considerable disadvantages soon became evident. For a start, and despite IKB's assurances to the contrary, atmospheric traction was costing the company twice as much as conventional steam power. This was because the pumping engines were running for longer than expected as they had to operate according to the timetable and keep pumping until the train had passed. Then serious problems began to emerge in maintaining the leather flaps which were vital in ensuring an airtight seal. The line runs right next to the sea and the flaps were constantly being soaked by the salty spray or dried out by the heat of the sun, and in the winter the flaps would freeze solid. In order to keep them supple several greasy coatings were tried, but these proved to be especially appetising to the local rat population who took to nibbling the leather, further reducing the efficiency of the seal and adding to the pumping costs.

In May 1848, a special committee reported to the board of directors that there was, 'an unexpected, rapid and continuing destruction going on in the leather' resulting in considerable expense in repairs. Embarrassed by the whole affair Brunel avoided contact with the directors, but in July they caught up with him at his office in Duke Street, London. He attempted to transfer all blame for the system's shortcomings to Clegg & Samuda who had given assurances that the valve would work correctly if properly maintained. The directors of the SDR had had enough, and from 10 September 1848 only conventional steam locomotives hauled the trains. The cost of the atmospheric fiasco to the directors and shareholders, including IKB, had been enormous and were estimated at almost half a million pounds.

Set between broad gauge rails, a long section of atmospheric pipe from the South Devon Railway on display at Didcot. The slot at the top was sealed by leather flaps hinged on one side, but these tended to dry out and smearing them with grease only attracted the rats. Several of the original pumping stations still remain along the Devon coast, including this one at Starcross with its truncated chimney.

This contemporary painting by Nicholas Condy shows the South Devon Railway at Dawlish, with the iron pipe running between the tracks and the pumping station, its chimney disguised as an Italianate tower, on the right. Today, the pumping station has gone but the railway, minus pipes, still separates this seaside town from the beach. (*Ironbridge Gorge Trust*)

Several of the engine houses still stand as monuments to what became known as the 'Atmospheric Caper'. At Starcross the red-stone building has lost part of its chimney and has been variously used as a church, a youth club and by a local sailing club. The one at Torquay is the most complete example with its chimney intact, and this is now occupied by various businesses, and the Totnes pumping station is occupied by Dairy Crest. But perhaps the most lasting legacy of the atmospheric railway is the dent it left in Brunel's reputation. No other project has attracted so much derision. The final word goes to the editor of *The Railway Times*:

> We do not take him for either a rogue or a fool but an enthusiast, blinded by the light of his own genius, an engineering knight-errant, always on the lookout for magic caves to be penetrated and enchanted rivers to be crossed, never so happy as when engaged 'regardless of cost' in conquering some, to ordinary mortals, impossibility.

To redress the balance, it should be noted that Brunel was not alone in putting his faith in atmospheric motive power. As mentioned, the first practical application had been the short railway at Dalkey which operated between 1844 and 1854. Clegg & Samuda also supplied equipment for the London & Croydon railway which ran for 7.5 miles (12 km) between London Bridge Station and Croydon, including a viaduct connecting Sydenham and the Crystal Palace. The first stage opened in July 1846, but the atmospheric system was abandoned the following year. Further afield, the Paris to Saint-Germain Railway ran a 5.3 mile (8.5 km) length of atmospheric track from 1847 until 1860. In New York the Beach Pneumatic Transit Company, which operated experimentally from 1870 to 1873, saw a new variation on the theme with the tube increased to sufficient diameter for the carriages to fit inside it as the piston. This principle was also used on the experimental Crystal Palace Atmospheric Railway of 1864 and in a similar fashion, although on a much smaller scale, unmanned pneumatic systems have been operated by various organisations all over the world to deliver goods or packages.

Brunel wasn't the only engineer to embrace atmospheric propulsion. In 1845, the London & Croydon Railway began tests with the Clegg-Samuda system on a 7.5 mile length of track. Three engine houses were built including this one at Croydon, above. However, the scheme was abandoned in 1847. Alfred Ely Beach turned the atmospheric propulsion on its head by making the carriage, or 'car', into the piston itself. Shown below is his demonstration of the system at the American Institute Fair in 1867.

Brunel's great lost bridge, the tubular railway bridge across the River Wye at Chepstow. This awkward asymmetrical site, with a high cliff on one side and lower ground on the other, called for a different approach. His solution was this 'closed' system dominated by the twin tubular box girders 300 feet long. Unfortunately, the bridge had become weakened by the 1950s and in 1962 it was replaced by an under-hung girder truss. In the image below, the A48 road bridge partly obscures the view of the rail bridge.

Chepstow's Tubular Bridge

Once described as 'hideously ugly', the 'Great Tubular Bridge' over the River Wye at Chepstow is arguably the least appreciated of all of Brunel's lost works. In my opinion this is a gross injustice and I shall make the case for putting the Chepstow Railway Bridge right up there, shoulder to shoulder, with his two greatest bridges, the Clifton Suspension Bridge over the Avon Gorge, and the Royal Albert Bridge at Saltash. I believe that Chepstow should be regarded not only as the missing link between the two methods of bridge building, but also as the dress-rehearsal that made the Royal Albert Bridge possible. And if you want hideously horrible, then take a look at the monstrosity that replaced Brunel's original at Chepstow.

The particular challenge that Brunel faced in getting a double rail track over the River Wye lay in the asymmetrical nature of the site. There are 120 feet (37 metres) high limestone cliffs on the eastern side and on the opposite shore an area of low ground comprised mostly of shingle and clay and rising only slightly above the high water level. The river is about 600 feet (183 metres) across at this point, has a wide tidal range and because it is navigable the Admiralty insisted on a vertical clearance of at least 50 feet (15.2 metres) at high tide, ruling out the possibility of a supporting arch of any sort. A conventional suspension bridge was also out of the question because while this type of bridge could cope with the loading of nineteenth-century road traffic, it was not suited to carry a substantially heavier railway train, weighing 100 tons or more. A suspension bridge would sag under the weight, creating a ripple effect as the train passed along it. What was needed was a far more rigid structure and the answer lay in the use of wrought iron.

Brunel had already built the wrought iron railway bridge over the Thames at Windsor, using three inverted bow-shaped girders from which vertical girders supported the decking. He had also dabbled with using wrought iron in a tubular or box girder form for the small swing-bridge in Bristol Docks. But it was his friend, Robert Stephenson, who had applied the box girder principle on a larger scale with his bridges for the Chester & Holyhead Railway at Conway, completed in 1848, and the larger Britannia Bridge over the Menai Straits in 1850. The Britannia Bridge consisted of four spans of 460 feet (140 metres) each in the form of huge box girders of wrought iron that were large enough to actually enclose the trains like a tubular tunnel. The advantage of using wrought iron in this fashion was that it has great tensile strength and the sections could be riveted together on site.

Abraham Darby's cast iron bridge across the River Severn at Ironbridge, Shropshire, was the first use this material in an arched bridge. (*Lance Bellers*) Brunel thought cast iron too weak and brittle for larger structures and preferred wrought iron which was becoming more readily available. In 1850, Robert Stephenson incorporated wrought iron box girders within the Britannia Bridge carrying the railway over the Menai Straits into Anglesey. But even Stephenson had considered a belt-and-braces approach initially, one which incorporated suspension chains, which is why the unnecessarily tall towers of the Britannia Bridge have square holes at the top.

This wrought iron swing bridge at Bristol Docks was Brunel's first experiment using tubular girders to form the top edges. Originally it was 120 feet 9 inches long and was positioned over the South Lock. Chepstow's railway bridge came next, but the ultimate expression of the tubular girder was the double-span of the Royal Albert Bridge over the River Tamar at Saltash. (*Network Rail*)

Two contrasting images of Chepstow Bridge: This contemporary illustration, published in *The Illustrated Magazine of Art*, *c.* 1853, is a thoroughly confused portrayal especially regarding the nature of the towers and the way the open-ended tubes seem to sit on flat lintels. Brunel would have been horrified. The tinted photograph from 1905 presents the bridge in a far more romantic, almost rose-tinted light. It does, however, partly show the three shorter spans on the western side. (*LoC*)

For the Chepstow Bridge Brunel opted for a novel belt-and-braces approach which combined the tubular girder principle with some aspects of a suspension bridge. On the eastern side of the bridge a 20-feet (6 metres) cutting led to a pair of towers pierced by arches and rising about 50 feet (15.2 metres) above the rail decking. At the other end of the 300 feet (91 metres) main span a similar pair of arches were supported on cylindrical iron columns, and beyond a further three 100-feet (30 metre) spans crossed the shore to meet a raised abutment. Between the pairs of towers there were two independent bridges side by side, each carrying a broad gauge track on a deck of plate girders. Suspension chains – adapted from the redundant chains for the Clifton Suspension Bridge – curved downwards from the towers while overhead two slightly arched tubes 9 feet (2.7 metres) in diameter counteracted the compression forces. Vertical struts and diagonal ties linked the deck to the tubular girders to ensure rigidity, although the deck girders were not fixed to the chains, but sat on rollers and saddles. In engineering terms the Chepstow Bridge is referred to as a closed system as all of the compression and tension is contained within the structure, as opposed to an open system on a conventional suspension bridge in which the loads are transferred by the chains to anchorage points on either side.

On 8 April 1852 the first of the 460-ton tubular girders was placed onto a pontoon consisting of six wrought-iron barges. Taking advantage of the high spring tide it was guided into position on the river, then lifted up to the level of the railway, and afterwards to its place between the towers. The first line of track was operational by July 1852 and the other was completed shortly afterwards the following year. *The Illustrated London News* observed that:

> The peculiarity of the site did not permit any display of Art – that is, of architectural embellishment; indeed, a pure taste rejects any attempt to decorate a large mechanical work with sham columns, pilasters, and small ornaments.

It appears that the Victorians weren't quite ready for the aesthetics of pure functionalism, although in truth the subtle shaping of the towers has something of the Egyptian flavour already displayed by Brunel on the Clifton Suspension Bridge. This understated Egyptian influence was also carried over to the towers on the Royal Albert Bridge at Saltash which is rightly regarded as Brunel's final masterpiece. The design of the Saltash bridge and, for that matter, its method of construction with tubular wrought iron girders assembled on the shore and then floated into position on pontoons, owe everything to Chepstow.

A tantalising glimpse of the bridge in this Edwardian postcard of the *Westward Ho* taking on passengers at Chepstow. This paddle steamer spent most of her life conducting excursions in the Bristol Channel. (*CMcC*) Below, another view of the site looking upstream, showing the iron support columns and the modern under-hung girder bridge.

Unfortunately, this story does not have a happy ending. By the 1950s, the Chepstow bridge was becoming too weak for the increasingly heavy rail traffic. In 1962, Brunel's original tubular structure was dismantled and an ugly underhung truss now supports the tracks. This is a prime example of the practical considerations winning over the intrinsic historical significance of this structure. I reassure myself that it couldn't happen today. Brunel's work is held in such high esteem that there would be a public outcry and the conservationists and engineers would have to find a way of strengthening the old bridge in a non-intrusive manner. Today thousands of motorists entering Wales on the A48 road bridge shoot over the River Wye without noticing the rail bridge running alongside. But if you take a little trouble it is worth pulling off the main road to get down to the western shore of the river to get a close-up view of what remains of Brunel's lost tubular bridge.

A contemporary engraving showing the railway crossing over the River Wye. Looking southwards it shows the high cliff on the eastern side of the river and the embankment on the low ground to the west.

Wooden Wonders

Although revered as a man of iron, IKB was not afraid to use other materials. This included timber when circumstances dictated, which meant, more often than not, when the budget was tight, and not just for his railway buildings – *as described in Railway Works* – but also for more substantial structures such as bridges and viaducts.

In 1841, Brunel constructed a wooden bridge to carry a public road over the railway cutting at Sonning on the eastern side of Reading. This two mile swathe cut through the landscape at depths varying from 20 feet (6 metres) to nearly 60 feet (18.3 metres) and at the level of the road crossing it was 240 feet (73 metres) wide. Once again J. C. Bourne provides us with an excellent depiction of the bridge which stood on four timber piers or trestles; two on either side of the track and another shorter pair halfway up each slope. The roadway consisted of a timber platform carried on three longitudinal beams which were supported by timber struts radiating like a fan from the top of the piers at about 12 feet (3.6 metres) below the road. There was a twofold advantage to this

method of construction using timber – it was relatively cheap as well as being quick to build – and Brunel would revisit this configuration for other timber viaducts, most notably in Devon and Cornwall.

Further down the line Brunel devised his first timber bridge to carry the railway itself, this time at an awkward skew angle over the River Avon immediately to the west of Bath Station. As we have seen, Brunel mistrusted cast iron as a material for bridges and under pressure from the Bristol Directors of the GWR to curb his overspending he came up with a unique design for a timber bridge. This had two spans, each consisting of six laminated ribs, built up of layers of wood bolted together, resting on masonry abutments and a single central pier. Iron ties connected the ends of the wooden ribs and the inner spandrels, the spaces between the arches and the platform, were criss-crossed by iron ties and braces. The open spaces on the outer ribs were filled in by ornamental cast ironwork. Bath's distinctive skew bridge was replaced by an ugly steel girder bridge in 1878, although the original masonry components still remain.

As the branch lines began to proliferate during the 1840s, Brunel emulated his design for the Sonning road bridge on the construction

Below: Bourne's lithograph of the wooden road bridge which crossed the Sonning Cutting.

Immediately to the west of Bath Station the railway crosses the Avon once more, this time at an awkward angle. Too sharp for a brick bridge Brunel devised a two-span skew bridge of laminated timber. Each 80 feet span consisted of six wooden ribs, in-filled with decorative ironwork. The masonry abutments and central pier still remain but the wooden bridge has been replaced by the trussed steel structure we see today.

of a number of wooden viaducts on the Swindon to Gloucester line, including nine in the seven mile stretch between Frampton and Stroud alone. For the Bourne viaduct, built across the Stroudwater Canal in 1842, the main span of 66 feet (20 metres) was supported by triangular trusses which rested in iron shoes on the top of masonry piers. On the St Mary's viaduct the main span was wider still at 74 feet (25.5 metres). That same year a timber viaduct was completed on the Bristol & Gloucester Railway at Stonehouse and this consisted of five timber trusses, 50 feet (15.2 metres) wide, standing on timber trestles. He also designed timber bridges for other railways including the Oxford, Worcester & Wolverhampton Railway, and the Birmingham and Oxford Junction.

Brunel built upon his experience by conducting a series of experiments to ascertain the strength of larger timbers and by exploring techniques for preserving the wood, in particular the Kyanising method which involved soaking the pre-cut timbers with perchloride of Mercury. This led to a far greater use of the material as the railway network spread across southern Wales and, later on, south-westwards into Devon and then Cornwall. The longest of Brunel's timber viaducts was at Landore, near Swansea. 1,760 feet (536 metres) long it had thirty-seven openings with a variety of spans ranging from the longest at 100 feet (30 metres) and down in increments to the smallest of around 40 feet (12.2 metres). The piers were of different materials, either masonry, timber or a combination of both, depending on the nature of the foundations. The main central truss consisted of a double polygonal frame connected by bolts and struts, and further strengthened by wrought iron tie-bars. A similar design was used at Newport, where there were only eleven spans, but this was destroyed by a fire before it had been finished and it was replaced by wrought iron trusses for the main span.

Back in England, the South Devon Railway had to pass over four deep valleys between Totnes and Plympton on the outskirts of Dartmoor. Four timber viaducts were constructed along the same design, the largest being at Ivybridge. This was on a curve and had eleven spans of 61 feet (18.6 metres) standing at its highest point at 104 feet (32 metres) above the valley floor. Designed to carry the atmospheric railway initially, the timber decking and framework was supported on pairs of slim masonry columns. When locomotives were introduced to this stretch of the line the viaduct was strengthened with an additional trussed parapet above the existing trusses.

Such was Brunel's confidence in timber construction that he even proposed a timber bridge for the difficult crossing over the river Tamar at Saltash. If this wooden wonder had ever been built it would have been the greatest timber bridge the world had ever seen, with six

spans of 100 feet (30 metres) and a central one of 250 feet (76 metres).
It is possible that IKB had taken his inspiration from his father, Marc,
who forty years previously had proposed a 'great bridge' with an 800
feet (244 metre) laminated timber arch to cross the River Neva at St
Petersburg. Fortunately for posterity's sake, if nothing else, IKB turned
to wrought iron to create what became one of his finest works, the
Royal Albert Bridge.

Work on the Cornwall Railway began in 1852, even though the Royal
Albert Bridge wouldn't be completed for another seven years. It was
particularly challenging countryside with a large number of deep valleys
to be crossed requiring the construction of forty-two viaducts between
Plymouth and Falmouth. For the most part these consist of masonry
piers 60 feet (18.3 metres) apart, centre to centre, rising up to within 35
feet (10.7 metres) of the decking. The piers were capped with iron plates
from which four sets of timber struts radiated like an upturned hand,
with further cross-bracing, to support longitudinal beams beneath the
deck. The tallest of this type of viaduct stood 153 feet (46.6 metres)
above the ground. On the West Cornwall Railway, where the height
was not so great, the spans were at 50 feet (15.2 metre) intervals and

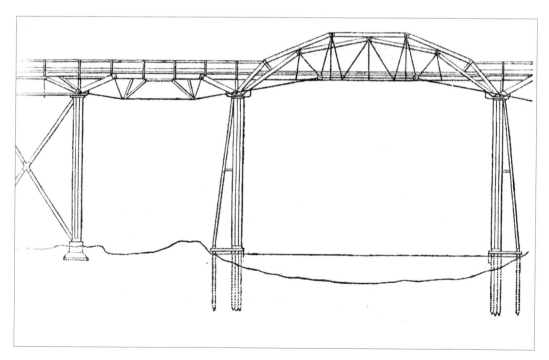

The longest of Brunel's timber viaducts was at Landore, crossing the Swansea valley in south
Wales. 1,760 feet long it had thirty-seven openings with a central span 100 feet wide.

The Coldrennick viaduct at Menheniot, *shown above*, is just to the north of the A38 at Lower Clicker. This was a typical example of a Cornish deep valley viaduct with tall buttressed masonry piers capped with a fan of Baltic pine struts supporting longitudinal beams beneath the deck. The Carvedras viaduct, *below*, located to the east of Truro railway station, was lower but still followed the same basic pattern, except with plain sided piers.

The curving sweep of the Moorswater in its prime. Once described as the most spectacular of Brunel's Cornish viaducts it was 147 feet high and 954 feet long with fourteen buttressed piers. Six of the original piers remain standing in the shadow of the replacement masonry viaduct which was completed in 1881. The remaining piers were dismantled as they were unsafe.

sat on timber piers. There were other variations to the structures. For example, in a shallow valley – such as the one at Pendalake to the east of Bodmin Road Station – the deck was supported on inclined timber legs sitting on small masonry piers. These piers were situated at 40 feet (12.2 metre) intervals and rose only slightly above ground-level. On the St Germans Viaduct the span was also 40 feet (12.2 metres) with support from a tapered timber and wrought-iron truss.

On the whole, the design of the timber components was standardised, partly to keep costs down, partly to facilitate easy replacement. Inevitably the timber would not last forever and during the lifetime of the viaducts special maintenance teams became adept at replacing individual struts. Because the price of Baltic pine continued to rise steadily, by the time the mainline track was doubled in 1908 the viaducts had all been replaced. (A few of the timber viaducts lasted on the branch lines until the 1930s.) In some cases the timber components were removed and the masonry piers were extended up to the full height of the deck. There is a good example of one of these at Menheniot, just eight miles into Cornwall along the A38 near Lower Clicker. Others did not fare so well and entire new viaducts were built alongside the old ones, leaving the unwanted piers standing like a row of tombstones.

Brunel has been described as 'the greatest timber engineer Britain has ever seen', but sadly not a single one of his timber bridges or viaducts has survived.

The replacement Ponsanooth viaduct under construction beside the old one. This was the tallest viaduct to the west of Truro and it became one of the last to be replaced, surviving until 1930. *Below*, the Coldrennick piers which were capped with brick extensions in 1898. During the work twelve workmen were killed in an accident when a platform suspended beneath the viaduct broke away. The extensions were strengthened by being encased in stone in 1933.

Military Designs

Brunel is best remembered as a civil engineer and he undertook very little work for the government. The reasons for this are twofold. First and foremost the main part of his career coincided with a sustained period of peace following the defeat of Napoleon's forces at the Battle of Waterloo in 1815. Secondly, he was used to getting things done at the pace and in the style he dictated and the ponderous and obstructive workings of the bureaucratic machinery of government did not fit well with his more impetuous nature. As a result there are very few examples of the great engineer applying his talents to the needs of war, the 'un-civil' side of engineering so to speak. It wasn't for any lack of patriotism, it was simply an overwhelming sense of frustration with government departments that systematically stifled innovation. He developed a particular scorn for the Admiralty Board as recorded in one of his letters from 1855:

> They have an unlimited supply of some negative principle which
> seems to absorb and eliminate everything that approaches them
> ... It is a curious and puzzling phenomenon, but in my experience
> it has always attended every contact with the Admiralty.

And when he wasn't firing a broadside at the Admiralty, it was another official body, the Patent Office, that thwarted his forays into military matters. Brunel never patented his own designs and condemned the patent laws as 'the curse of the day, and the source of greatest injury to inventors and manufacturers, and still more to the public.' This view was the result of his experience with a design for a polygonal bored rifle. In October 1852 he had contacted the gunsmith Westley Richards, of Birmingham, with a view to having an experimental barrel made for him.

> 'I want a rifle barrel made octagon shaped inside, the octagon
> having a twist rather more than usual, and an increasing twist,
> say twice as much at the mouth of the piece as at the breech.'

The rifle barrel was made and apparently it performed well in trials, however Brunel's hopes of taking it further were thwarted when Jospeh Whitworth obtained a patent for just such a design in 1855. This must have infuriated Brunel who was now required to obtain a license from Whitworth if he wished to continue with his work and, consequently, he abandoned the project.

'A quiet day in the diamond battery – portrait of a Lancaster 68 pounder, 15 December 1854.' The Crimean War saw a bewildering mix of old style armies, huge siege guns, plus the first wartime use of various new technologies including steamships, railways and telegraphy. Although Brunel's 1855 designs for a flat topped semi-submersible siege-gun were ignored by the British Admiralty, they were echoed by the US Navy's ironclads such as the USS *Monitor*, shown below, which was launched in 1862. (*LoC*)

In March 1854 Britain and France declared war upon Russia in support of Turkey and Sardinia who had been fighting in the Crimean Peninsula since the previous autumn. The Crimean War is sometimes referred to as the first 'modern' war as it saw the first use of steamships, railways, the electric telegraph and the pioneering art of war photography. In most other respects it was a thoroughly outmoded war typified by logistical and tactical errors throughout the land campaign. The British soldiers who sailed to the Crimea on a rag-bag collection of ships, including the *Great Britain*, were organised and equipped in ways that had little changed since the Napoleonic wars fifty years earlier. Like many of his contemporaries, Brunel was keen to apply his skills to the needs of the nation and he turned his attention to large siege guns, or cannon, in particular the problem of preventing their barrels from fracturing or bursting. One method he proposed was to wrap the barrel with iron wire to make it more resistant to explosion, but once again a prior patent put paid to that line of experimentation.

Brunel's next attempt to aid the war effort brought him up against the Admiralty once again. In 1855, he devised a floating or semi-submersible siege gun which, he believed, could be deployed effectively against the Russian fortresses on the Baltic Coast. His design resembled a submarine with a flat top and only the hemispherical turret housing a 12-inch gun protruding above the water line. This would fire at a rate of three rounds per minute with the ammunition breech-loaded mechanically from within the hull. As the vessel was required to operate in shallower waters close to the shore, Brunel dispensed with a propeller and instead he proposed an innovative propulsion system using water jets. A main jet at the rear would propel the vessel forwards and smaller ones on either side gave lateral control. These would also provide the means of aiming the gun laterally with the gunner elevating the gun as required. Because the jets only gave short-term propulsion he devised a larger steam-driven vessel – a sort of 'mother-ship' – which would carry the floating siege gun within its bows and release it, via doors, at the location where it was to be deployed. In effect this concept anticipated the principle of the landing craft which would play such a decisive role almost a century later on the D-Day beaches.

IKB duly submitted his designs together with a scale-model of the floating siege gun to the Admiralty which responded with customary indifference. In fairness, Brunel's design had brought nothing new to the table apart from the novel propulsion system. What's more, he wasn't the only engineer, inventor or crackpot pestering the Admiralty with their ideas. Prince Albert himself had taken an interest in the

'The Lady with the Lamp' making her rounds at the notorious Scutari hospital during the Crimean War. Florence Nightingale worked tirelessly to improve conditions in the over-crowded hospitals, where more men died from infections and disease than on the battlefield, and also to raise awareness of their plight. Roger Fenton was one of the first war photographers and in 1855 he took this picture of Balaklava, below, with the hospital shown on the right. (*LoC*)

gunboats and even John Scott Russell – the man who was building the *Great Eastern* at the time and who has often been portrayed as Brunel's nemesis – had constructed a full-sized gunship at his shipyard. It is interesting to note that during the American Civil War, which began only five years after the Crimean War had ended, the design of the US Navy's first ironclad, the USS *Monitor*, closely resembled that of the semi-submersible floating siege gun, albeit considerably bigger and with a fully rotating armoured gun turret.

Renkioi Hospital

An unexpected aspect of the Crimean War was the emergence of the war correspondent, thanks primarily to the immediacy of the electric telegraph. William H. Russell's front-line reports in *The Times*, aided by Roger Fenton's photographs, gave the British public its first sense of the reality of war. Not just the horrific loss of life, but also the terrible conditions being endured by the soldiers; their own sons and husbands. Fighting a war at such a distance had pushed the supply chain beyond breaking point. Essential provisions simply weren't getting through and the care of the wounded and sick was woefully inadequate if not downright scandalous. It was Florence Nightingale who highlighted their plight, particularly at the British hospital established at the former army barracks at Scutari. Conditions in the overcrowded and understaffed hospitals were truly appalling. Treatment was minimal and cholera, dysentery and typhoid were rife, resulting in the deaths of more men than died on the battlefield. Just as with Brunel, Florence Nightingale was enraged by the intransigence of the bureaucrats in Westminster, most notably the permanent Under-Secretary of War, Sir Benjamin Hawes, who she described as 'a dictator, an autocrat irresponsible to parliament, quite unassailable from any quarter, immoveable in the midst of so-called constitutional government.'

Sir Benjamin Hawes happened to be the friend and brother-in-law of IKB, and whether or not he deliberately set out to snub 'that tiresome woman', he turned to Brunel for help. On 16 February 1855, the War Office commissioned him to design a temporary hospital, one that could be prefabricated in sections and shipped to the Crimea for erection on any reasonably level piece of ground. Brunel's response was typical; highly detailed and put into practice in an incredibly short time. It was as if he was putting on a display, showing the bureaucrats how you get a job done when there is the will.

It is only by the prompt and independent actions of a single individual entrusted with such powers that expedition can be secured and vexations and mischievous delays avoided... These buildings, if wanted at all, must be wanted before they can possibly arrive.

His team had the plans for the hospital drawn up within days and by early February a test version of one of the wards had been erected at Paddington for evaluation.

To be constructed in wood, the layout of the hospital consisted of a central corridor or spine with the ward buildings, or units, pointing outwards from either side, with a wide gap between them to allow the air to circulate. Each building unit was designed to contain two wards with twenty-five beds in each, a nurse's room, small store room, bathroom, surgery plus proper lavatories. The pitched roofs allowed for air to circulate and they were clad with polished tin to reflect the heat of the summer sun. Long narrow windows immediately under the eaves let in the daylight without the glare of direct sunlight. The walls were insulated between the cladding and internal boards to hold the heat within the building during the winter and outside during the summer. Fresh air was pumped through vents from under the floor, and there were opening windows in the eaves and gable ends. Each unit linked into a system of main drains made of wooden trunking and Brunel's attention to detail was meticulous. His notebooks include various sketches for the water closets, wash basins and so on.

Once completed the pre-fab hospital was taken to Turkey on twenty-three ships. Put in charge of the on-site construction team was John Bruton and together with Dr Edmund Parkes, the Medical Superintendent of the new hospital, they identified a suitable site at Renkioi in the Dardanelles. The hospital was manufactured, shipped and assembled ready to receive some patients within five months, and the full 1,000 capacity by 4 December 1855. It had been an incredible achievement, but the war was over in January 1856 and most of the patients were sent home by May. No longer needed, the hospital was sold off piecemeal with some of it providing housing in the area.

Of the 1,331 patients who were treated at Renkioi, only fifty died – a fatality rate of 4 per cent. By comparison the fatality rate at the Scrutari hospital had been a staggering 42 per cent, although it should be pointed out that Renkioi's distance from the fighting meant that the majority of its patients were convalescing rather than the recently wounded.

Brunel didn't invent the concept of prefabricated construction. After all Joseph Paxton's Crystal Palace for the Great Exhibition of 1851 had been built in this manner. What he did do so effectively was demonstrate that the principle could be appropriately applied to a wider range of functional buildings, from hospitals and homes to factories and industrial buildings.

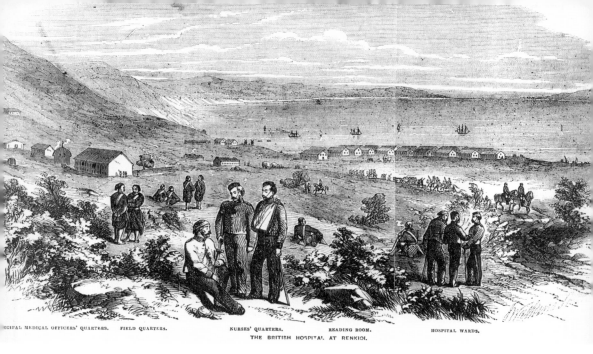

CIPAL MEDICAL OFFICERS' QUARTERS.   FIELD QUARTERS.   NURSES' QUARTERS.   READING ROOM.   HOSPITAL WARDS.

THE BRITISH HOSPITAL AT RENKIOI.

In February 1855 the War Office commissioned Brunel to design a temporary hospital which could be prefabricated and shipped out to the Crimea. Brunel worked fast and by December that same year the hospital at Renkioi was ready to receive its full capacity of 1,000 patients. It is shown in this engraving from *The Illustrated Times*, with the main hospital wards to the right. The elevation and plan view show one of the wards with the central corridor on the left.

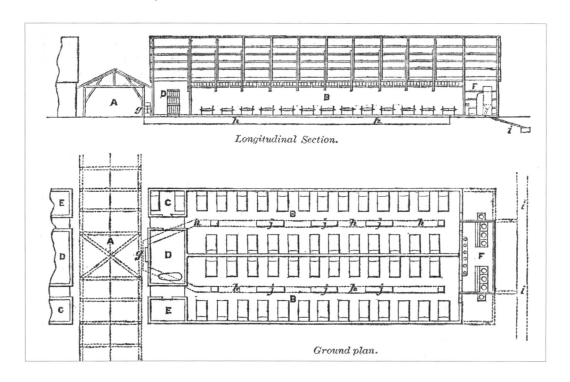

Longitudinal Section.

Ground plan.

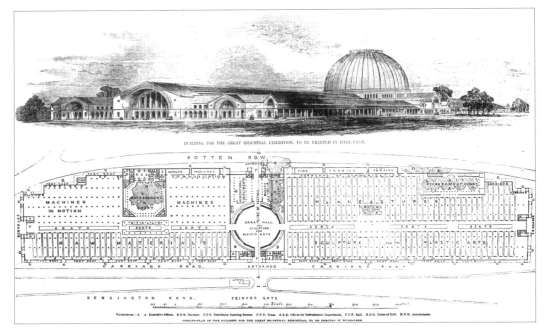

BUILDING FOR THE GREAT INDUSTRIAL EXHIBITION, TO BE ERECTED IN HYDE PARK.

GROUND-PLAN OF THE BUILDING FOR THE GREAT INDUSTRIAL EXHIBITION, TO BE ERECTED IN HYDE-PARK.

In 1850, Brunel sat on the committee tasked with finding a design for a temporary building to house the Great Exhibition in Hyde Park. The committee came up with its own collective effort, above, and his contribution was the 150ft high iron and glass dome. The design was not popular and fortunately Joseph Paxton came up with a conservatory-inspired glasshouse christened by *Punch* as the 'Crystal Palace'. After the exhibition had finished it was rebuilt in an enlarged form at Sydenham, *below*.

The Two Towers

IKB's involvement with the Crystal Palace is a long and convoluted story that encompasses a massive dome that was never built, revolutionary new methods of construction using iron and glass, and a pair of towers designed by him that survived a terrible fire only to be deliberately destroyed by the authorities during the Second World War.

The inspiration for the Great Exhibition of 1851 grew out of the desire to emulate the French Industrial Exposition of 1844 and to reassert Britain's prominence as an industrial powerhouse. The main driving force behind its instigation was Henry Cole of the Society of Arts, Manufacturers and Commerce who recruited the support of Prince Albert who, as the Queen's Consort, was grateful to have a public role of his own. In 1850, a Royal Commission for the exhibition was established with the Prince as chairman, and a building committee was appointed with members drawn from the aristocracy and the worlds of architecture and engineering, the latter including Robert Stephenson, William Cubitt and Isambard Kingdom Brunel. A competition was held for the design of a temporary exhibition building to be erected in Hyde Park, and 230 entries were reviewed by the committee. As none met the budget of £100,000, or the requirement for completion by May the following year in time for the opening of the exhibition, the committee set about producing its own design. The result was the proverbial designed-by-committee 'camel' of a building; three brick-walled naves – looking not unlike a railway shed – capped by Brunel's contribution, a dome of iron and glass that looked like a vast cloche 150 feet (46 metres) high and 200 feet (61 metres) in diameter. If ever a building deserved to be derided as an 'unsightly carbuncle' then it was this one.

Fortunately, it was at this point that Joseph Paxton took an interest. Head gardener at the Duke of Devonshire's Chatsworth House in the Peak District, Paxton was entirely self-taught. Between 1837 and 1840 he had built the 'Great Conservatory' at Chatsworth, a huge glasshouse 300 feet (91.5 metres) long, using an innovative combination of cast-iron and glass. Despite the fact that it was known less politely as the 'Great Stove', because of its appetite for vast quantities of coal to heat it, Paxton expanded the conservatory principle to create a building for the Great Exhibition. A vast greenhouse, the biggest the world had ever seen, 1,848 feet (563 metres) long. Appropriately christened by *Punch* magazine as the 'Crystal Palace', it was constructed from prefabricated sections and completed in less than five months in time for the opening

SOUVENIR OF THE CRYSTAL PALACE
DESTROYED BY FIRE NOV. 30TH 1936

SORROWS OF A SPONSOR

MR. PUNCH. "AND TO THINK THAT IT WAS I THAT GAVE YOU YOUR NAME WHEN I WAS A MERE LAD!"

[In 1850, before the Great Exhibition began, Mr. Punch jokingly applied to the Crystal Palace the title which throughout its whole existence it continued to bear.]

On the night of 30 November 1936 the Crystal Palace was destroyed by fire. The heat was so intense that it sent a plume of steam up through the chimney of one of Brunel's towers, shown in the postcard, *left*. As he surveys the ruins Mr Punch laments, 'To think that it was I that gave you your name.' But at least the two towers had survived the blaze. (*CMcC*)

ceremony on 1 May 1851, which IKB and his family attended. The exhibition was a great success and Paxton's structure was a glittering triumph, literally. It certainly has many parallels with IKB's designs for the new Paddington Station – *see 'The First Paddington'* – and Brunel's associations with the Crystal Palace didn't end there.

As Hyde Park had only been intended as a temporary venue for the Great Exhibition, once it finished in October 1851 there was the question about what to do with it. The Crystal Palace Company was formed and one of its directors happened to be John Scott Russell, an old acquaintance of Brunel's and the shipbuilder for the *Great Eastern* just a few years later. A new home was found for the Crystal Palace on Sydenham Hill, south London, where it was rebuilt between 1851 and 1854 in an extensively altered and enlarged form, shorter than the original but much higher. Sir Joseph Paxton, as he now was, called upon Brunel to help with the building's reconstruction on the new site.

The parkland in which it stood at Sydenham was laid out as formal gardens by Paxton featuring grand terraces and water features including cascades and elaborate fountains, with plumes up to 250 feet (76 metres) high, which obviously needed a copious supply of water at pressure. This would be supplied by a pair of water towers, one at either end of the building and separated slightly from it, which would also feed water to the boilers heating the building and contain their chimneys. The first designs for the two towers was prepared by Paxton's assistance, Charles Hurd Wild, and when Paxton became anxious about what he saw he consulted IKB on the matter. Brunel, in customary manner, pulled Wild's designs apart:

> The attempt to support upwards of up to 500 tons at a height of more than 200 feet [61 metres] upon a cluster of slender legs with but a small base involves considerable difficulties.

He also pointed out that the main legs of the water should not double up as water pipes, it was necessary to provide horizontal bracing to prevent the legs from buckling, the water tank should be of wrought iron not cast iron, and the weight of the tank should be carried on the top of the columns and not through supporting struts. The foundations also needed beefing up. Consequently Wild was swiftly ousted in favour of Brunel and Wild's half-built towers were dismantled. With Brunel on board as engineer, Paxton upped the water capacity of the tanks from 500 to 1,500 tons. The foundations had a concrete ring with a diameter of 58 feet (17.7 metres) capped with slightly tapered brickwork up to a height of 18 feet (5.5 metres). On this stood the cast-iron plates from which rose twelve cast iron columns. Externally the infill panels between the columns echoed the design on the main building, and inside was a brick chimney with an iron staircase spiralling around it.

After several delays the towers were finally ready by the summer of 1856 and they began to deliver water when the fountains were turned on in the presence of the Queen. She was back again the following year to perform the second official opening ceremony for the Crystal Palace. The new venue soon settled into a routine of concerts, exhibitions and public entertainments, but by the end of the century the novelty was beginning to wear thin and the building was falling into disrepair. By 1911 it was no longer paying its way and in 1913 it was purchased by the Earl of Plymouth and subsequently re-purchased for the nation when funds were raised by public subscription. After a spell as a naval training establishment during the First World War, the Crystal Palace was gradually restored as a public venue with the north tower opened to sightseers with a taste for heights. In 1933, the television pioneer John Logie Baird came to Sydenham where he attached transmission aerials to the south tower and also built the first television studios in the main building. Then, on the night of 30 November 1936, disaster struck and the Crystal Palace was obliterated in a fire. It is estimated that 100,000 people saw the blaze, among them Winston Churchill who observed, 'This is the end of an age.' Miraculously both of Brunel's towers still stood tall and proud the following morning, and they remained prominent landmarks in the area for several years. Perhaps too prominent.

The official reason for getting rid of the towers is that they might have served as navigational aids for the Luftwaffe bombers raiding London. In 1941, the north tower was demolished with a 220 lb charge of explosives. Against a sky peppered with barrage balloons it fell as if in slow motion, a final plume of soot and dirt billowing from the top of the internal chimney. The south tower, because of its proximity to other buildings, couldn't be felled in such dramatic fashion and it had to be dismantled piece by piece. All that remains of the two towers are the foundations, and of the main building itself there is the terracing and a handful of statues. There is also a small museum near the site of the south tower.

The fall of the north tower. With the coming of war it was decided that the prominent towers might serve as navigational aids to Luftwaffe pilots looking for London. In 1941, the north tower was toppled by an explosive charge and it landed on some of the outlying buildings, including the former aquarium, which had survived the 1936 fire. The south tower, because of its proximity to other buildings, was dismantled. (*CMcC*)

The Great Ship

Everything about Brunel's third and final ship, the *Great Eastern*, was big. Years ahead of her time, when launched in 1858 she was larger than any ship afloat; a record that would last for almost half a century. Her vital statistics were impressive: 692 feet (211 metres) long – more than twice IKB's previous iron ship, the *Great Britain*, and almost three times longer than the wooden-hulled *Great Western*. Intended to carry up to 4,000 passengers, she sailed with only thirty-five on board on the first voyage of her all too brief four-year career as an ocean liner and only 100 on the next. This colossal ship took five years to build and almost three months to launch. Little wonder then that the *Great Eastern* caused her creator so much grief towards the end of his career and his life.

One unexpected outcome of the international flavour of the Great Exhibition of 1851 was an increased interest in the wealth and resources of other countries, including India and the USA. In March 1852, Brunel sketched a rough design in his notebook for a massive steamship, to carry emigrants and goods, under the heading 'East India Steamship'. He shared his thoughts with his friend John Scott Russell, an experienced naval architect and ship builder, and Russell suggested they approach the Eastern Steam Navigation Company which had been formed a year earlier to exploit this increased trade, not to India but across the Atlantic. However, when the government awarded a vital lucrative mail contract to the rival Peninsular & Oriental Steam Navigation Company, Brunel turned to his friends to raise the money to build the ship, and he also took a large financial stake in the company himself:

> By December 1853 after great exertion 40,000 shares were taken and £120,000 paid up ... More than once we have nearly failed and broken down ... After two years' exertions we are thus set going, contracts entered into and work commenced 25 February 1854.

Brunel was appointed as engineer to the project while the actual construction of the ship was put out to tender, a process won by John

Opposite: These hefty timbers are the remains of the wooden slipway on which the *Great Eastern* was constructed at the Scott Russell shipyard in Millwall.

Scott Russell no less, who offered a very competitive price for building the ship provided a second one was ordered once it was completed.

In comparison with the graceful curves of the *Great Britain* the *Great Eastern* appeared to be a brutal statement of intent, her hull was slab sided with an almost vertical prow, and she had a flat upper deck without the superstructures we associate with modern ocean liners. Work began on the ship in the spring of 1854 beside the Thames at Millwall, on the Isle of Dogs, where she grew piece by piece until she loomed above the surrounding ramshackle buildings. Completing the 'Great Ship', proved to be an epic struggle between Brunel and Russell. IKB's celebrated biographer L. T. C. Rolt has portrayed their working relationship as a battle of wills and egos and he cast Russell as the villain of the piece. But it might be fairer to say that this was an epic struggle of Brunel's ego and his need to dominate any project.

Below, laying the transatlantic telegraph cable in 1865-66. Her fourth funnel was removed to make space for cable machinery. In her career she laid over 32,000 miles of telegraph cable.

A notoriously demanding and difficult man to work for or with, IKB was incapable of delegating any responsibility to the long-suffering Russell as he saw it as detracting from his own central role in the project. Admittedly the overall concept had been Brunel's, but his constant interference and manipulation exasperated the ship-builder and probably delayed the completion of the *Great Eastern* by at least a year, maybe longer.

In designing such a big ship IKB had once again put great emphasis on its strength. This was achieved with a double skinned, or rather double bottomed hull with the iron plates set 2 feet 10 inches (86 cm) apart and containing longitudinal members spaced at 6 feet (1.8 metre) intervals. Brunel described the overall structure:

Below, when it came to launching the great ship in November 1857, the wooden cradles remained stuck fast to the slipway. *Overleaf*, a contemporary diagram of the ship in elevation and plan form, with the prow to the right. Note the two engine rooms, one to drive the paddles and another further back for the propellers.

LONGITUDINAL SECTION

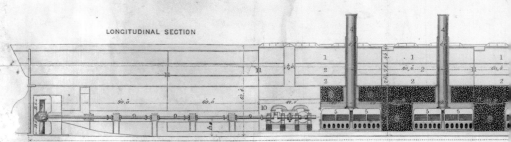

PLAN

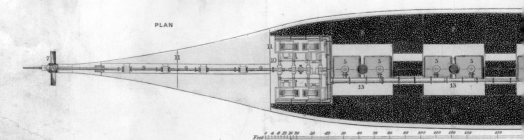

Feet

THE GREA

DESIGNED BY I. K.

This immense vessel is the property of the Eastern Steam Navigation Company, incorporated by Royal Charter. The capital of the Company is $6,000,000, in $100 shares, with power to increase the amount to $10,000,000.

The Great Eastern was designed by Isambard Kingdom Brunel, Esq., F. R. S. The ship and paddle engines are building by Messrs. Scott, Russell & Co., at Millwall, on the river Thames; and the screw engines by Messrs. James Watt & Co., Soho Works, Birmingham.

The principal dimensions of the ship, her capacity and power, are as follow:

Length (rather more than the eighth of a mile),	680 feet.	Diameter of cylinders,			74 inches.
Breadth,	83 "	Length of stroke,			14 ft. 6 inches.
Depth from deck to keel,	60 "	Draft of water (laden),			30 feet.
Length of principal saloons,	400 "	" " (light),			20 "
Number of Decks,	4	To carry {800 1st class}			
Tonnage,	22,500 tons.	" {2,000 2d " }			Total, 4,000 Passengers.
To carry coals and cargo,	18,000 "	" {1,200 3d " }			
Nominal power of paddle engines,	1,000 horses.	" Troops, without other passengers,			10,000.
" " screw	1,600 "	Weight of iron used in construction, about			7,000 tons.
Number of cylinders of paddle engines,	4				

The speed of the vessel is estimated by Mr. Brunel at fifteen knots an hour, without diminution and without cessation, under any weather; a speed which would accomplish the voyage between England and India, by the Cape, in from thirty to thirty-three days; and between England and Australia, in from thirty-three to thirty-six days.

The mode proposed for launching the vessel is as follows: In constructing the foundation of the floor on which the ship is being built, provision is made at two points to insure sufficient strength to bear the whole weight of the ship when completed. At these two points, when the launching has to be effected, two cradles will be introduced, and the entire fabric will be lowered down gradually to low-water mark, whence, on the ensuing tide, the vessel will be floated off.

One peculiarity of this ship is that her deck will be flush, except for cabin entrances and similar purposes, so that her great length will afford the passengers a promenade of more than a quarter of a mile round the deck, which, from the magnitude of the vessel, ought to be at all times free from shipping water.

Intimately connected with the appearance of the . . . chances of accident been so much studied as in the C . . . 1, An inner and outer skin in compartments; 2, Wa . . . propeller.

With regard to motive power, the wind will be us . . . The centre masts will be the principal, these will be . . . size. The sails will be of an effective but simple char . . .

It is, however, in respect of its steam power, that . . . wheel and the screw. The engines are incomparably . . . far greater than their nominal power already stated. . . . each other. The vessel will have ten boilers, and fif . . . desired. The boilers will be placed longitudinally alo . . . when it is stated that every boiler will have ten furn . . . be used will be anthracite coal.

The paddle engines are direct, acting with oscillati . . . they may be used jointly or separately, so that both . . . diameter of the paddle-wheels will be sixty feet.

There are few points connected with a paddle-whe . . . less, in many cases, altogether changing the characte . . . of this ship will show the real difficulty in the way of . . .

The vessel will draw ten feet less when light than . . . wheel is, in itself, an important consideration.

The screw propeller will be twenty-four feet in d . . . manner.

The best terms to describe the build is by stating . . . skin, or ship, is two feet ten inches. These skins are . . .

NEW YORK

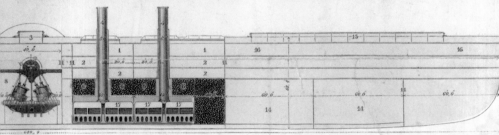

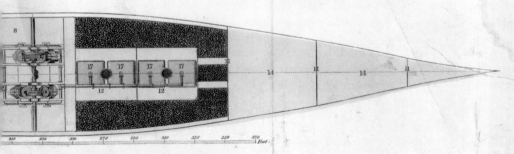

EASTERN.

SQ., F. R. S., &c., &c.

ded thereto, her motive power. In no ship have the
vision against such contingencies may be thus stated:
3, Ample masts and sails; 4, Paddle-wheels; 5, Screw

s purpose the vessel will be provided with seven masts.
a line-of-battle ship; the other masts will be smaller in

distinguished, combining, as it does, both the paddle-
made for marine purposes; and their *actual power* will be
ifferent parts of the ship, and be entirely independent of
er can be cut off from its neighbor, and used or not, as
; and it will give some idea of their generative power,
whole no less than one hundred furnaces. The fuel to

nstructed on the disconnecting principle, in order that
dle-wheels, can be put in independent motion. The

ant than the diameter of the wheel ; six inches, more or
A consideration of the light and heavy draft of water

ll the great lift of water at the deepest immersion of the

ced at the stern of the vessel, and worked in the usual

d outer skin. The space between the outer and inner
agitudinal webs or girders formed of plate and angle iron.

EL D. BRAIN.

There are 17 of these webs on each side of the ship, which run the entire length of the vessel ; and they are placed at such distances as to
extend upwards at intervals of about three feet, from the keel to the main deck ; and they are again closed up in lengths varying
from twenty to sixty feet. Thus the outer and the inner ships are joined together by means of a great number of water-tight webs or cells
of extraordinary strength, giving the vessel a rigidity such as has never been communicated to a ship before. The upper deck is
treated in the same manner for a width of twenty feet on each side, and iron girders bind one side to the other, so that the entire
vessel becomes, as it were, a beam of strength, and the whole fabric may be denominated a web of woven iron, the rivets forming
the fastenings, and the webbed or honey-combed cells becoming an indissoluble structure. The web plates are of half-inch iron, and
the outer and inner skins are of three-quarter inch iron. The compartments between the outer and inner skin will hold 3,000 tons
of water ballast, should it be required. The floor of the ship, as previously stated, is perfectly flat, the keel being turned inwards,
and riveted to the inner ship's keel. The bow and stern have additional strength imparted to them by strong iron decks at those
parts.

The ship will have a number of large ports on the lower deck to receive railway waggons, carriages, and bulky goods. She has
also sixty ports on each side, two feet six inches square, for ventilation ; and an abundance of dead lights. The lower ports are ten
feet above the water when the ship is loaded.

In addition to these safeguards outwardly, the vessel is divided transversely by ten separate water-tight bulk heads running up
to the main deck, and these again are crossed by longitudinal bulk heads running fore and aft.

It may therefore be said, that the ship consists inwardly of a great number of small cells, or water-tight compartments, between
the outer and inner skins, and of a number of large square compartments in the body of the vessel. The cabins will be on the decks
above these compartments, and will form large and splendid saloons. The captain and officers' berths will be placed on the upper
deck.

REFERENCES.

1. Upper Saloons.
2. Principal Saloons.
3. Captain's Room.
4. Funnels.
5. Boilers for Screw Engines.
6. Coal Bunkers.
7. Four-bladed Screw.
8. Paddle Engines.
9. Screw Shaft.
10. Screw Engines.
11. Cross Bulk-heads.
12. Steam Pipe from forward Boilers.
13. Steam Pipe from after Boilers.
14. Space for Cargo.
15. Officers' Rooms.
16. Berths for Crew.
17. Boilers for Paddle Engines.

'The whole of the vessel is divided transversely into ten perfectly water-tight compartments, by bulkheads carried up to the upper deck. So that if the ship were to be cut into two, the separate portions would float.'

Combined with the upper deck, also of cellular construction, and two 36 feet-deep longitudinal bulkheads, the result was a huge box girder.

The expected method of construction would have been within a dry dock which would have been flooded once the hull was completed, but with such an immense ship creating a big enough dry dock would have been very expensive. Brunel opted for a sloping slipway constructed of massive oak timbers resting on piles. The ship's hull sat on a cradle and this would slide down the sloping timbers into the river, held in check by chain drums. Similar sideways launches had been tried before in the USA, but never on this scale and always as a free or un-restrained launch.

The day for the launch was set as 3 November 1857, a date dictated by the tides. Much to Brunel's dismay the company directors had sold tickets to thousands of rowdy onlookers and a grey drizzle failed to dampen their boisterous barracking. Brunel conducted the proceeding with a system of signals and when the time came for the steam winches to start there arose a thunderous noise as the chains – so familiar from Robert Howlett's photograph of Brunel – reverberated against the hollow iron hull. The cheers of the crowd were cut short when suddenly the bow jerked forward, causing a wooden lever on one of the checking drums to flail out of control, injuring several of the workmen and killing one of them. The launch was halted. On the next attempt, on 19 November, the ship did move, but only by about 14 feet (4.3 metres). Only brute force could overcome the friction of the cradle against the wooden slipway and huge hydraulic jacks were brought in to shove the ship, inch by painfully slow inch, into the Thames on 31 January 1858.

By early September in 1859 the *Great Eastern* had been fitted out and was finally ready for her maiden voyage. A couple of days before she was due to sail Brunel came on board to inspect the ship. He posed for a photograph standing beside one of the tall funnels. Resting on a stick, and with his stovepipe hat removed, he looked old and worn out. Moments afterwards he slumped to the deck, some say as the result of a heart attack. The ship sailed without him and as it proceeded down the English Channel, just off the coast from Weymouth, it was rocked by a terrific explosion. A valve on a feed-water heater at the base of a forward funnel had been left closed causing the explosion and a deadly blowback of steam that claimed the lives of five stokers. Lying

In September 1859, just days before the *Great Eastern* was due make her maiden voyage, Brunel came to the ship and was photographed standing against one of the funnels. He was very ill and moments later he collapsed. The ship sailed without him, and on 8 September it was rocked by a powerful blast. A valve on a feed-water heater at the base of a funnel had caused the explosion and the blow-back of steam scalded five stokers to death. Brunel died a week later. The damaged funnel is shown above.

The *Great Eastern* was decades ahead of her time, not just in terms of her vast size, and she was very big at 692 feet long, but also because of the way she had been constructed. Unfortunately there wasn't the market for such a large passenger vessel and she had a more successful second career as a cable laying ship. In 1865, she laid the first transatlantic telegraph cable. The *Great Eastern* ended her days on the Mersey, serving as a showboat and giant advertising hoarding. In 1888, she was sold for scrap and is shown, *below*, awaiting her fate.

in his bed at his Duke Street home, Brunel was distraught when he received the news. He died on 15 September 1859. The myth-makers would have us believe that it was the *Great Eastern* that killed Brunel. Although the stresses in building and launching the great ship might have hastened it, the actual cause of death was more prosaic: A kidney condition referred to as Brights disease in Victorian times and now known as chronic nephritis.

Starting in 1860, the *Great Eastern* made only nine transatlantic round trips, including a charter by the British government to transport over 2,000 troops to Quebec in 1861. However, with the shipping company making heavy losses, the ship was put up for auction on 14 January 1864. The start price was £50,000 – a fraction of her build cost – but as there were no bids and she was withdrawn from sale. Three weeks later the auctioneers tried again, this time without a reserve, and the ship sold for a mere £20,000 to a group of businessmen including Daniel Gooch. They set up the Great Eastern Steamship Company and chartered the ship to the Telegraph Construction & Maintenance Company. Once refitted for her new role the *Great Eastern* successfully laid the first transatlantic telegraph cable in 1865, on the second attempt, and over the next few years went on to lay cables from Brest, in France, to Saint Pierre and Miquelon, and also between Aden and Bombay.

At the conclusion of this very successful period the end of the *Great Eastern* came in stages. For several years she served as a showboat, a floating concert hall and gymnasium, sailing up and down the Mersey as an advertising hoarding for Lewis's Department Store. Then in 1888 she was sold as scrap. But the old ship had one last trick to play. So strongly was she constructed that it took nearly two years for the workmen to pull her apart. Even in her death throes, it was as if the ship's vast scale and the audacity of her creator were too big for the world to cope with.

Brunel submitted several entries in the competition to design a bridge at Clifton. The final version, his 'Egyptian thing', above, featured Sphinxes gazing across the Avon Gorge, although other versions show them pointing outwards. The bridge was only completed after his death and in a substantially modified form. This view, *below*, is from the Leigh Woods side looking across to Clifton and the old observatory on the Downs. The underslung cradle is moveable and provides access for the maintenance crews.

Clifton Suspension Bridge

At first sight the Clifton Suspension Bridge might seem an unlikely subject for a book on the lost works of Isambard Kingdom Brunel, after all it still spans the Avon Gorge at Bristol for all the world to see. Even so, the Clifton Bridge earns its entry on three counts. Firstly, Brunel produced drawings for several alternative designs which were never built. Secondly, the bridge was not completed during his lifetime – which explains its appearance at the end of this book – and thirdly, the bridge that was finally built differs from Brunel's design in several significant respects.

Following his departure from the Thames Tunnel project following the deluge in 1828, Clifton Bridge was IKB's first major commission. It launched his meteoric career and cemented his relationship with Bristol, the city that became most closely associated with his work. The first proposals for a crossing over the Avon Gorge had actually appeared some fifty years or so before Brunel was born. In 1754, a prosperous wine merchant named William Vick got the ball rolling by bequeathing the sum of £1,000 to be invested until it grew to £10,000 when it was to be used for the construction of a stone bridge. By the turn of the century Clifton had become a fashionable place to live for Bristol's wealthy merchants and the appropriately named William Bridges published his vision of a grand stone-built bridge. This extraordinary structure would have filled the gorge like a curtain, its five storeys containing houses, a granary, corn exchange, chapel and tavern. Surmounted by a lighthouse, it was more like a vertical village than a bridge and inevitably this impossibly expensive design was never realised.

By 1829, Vick's legacy had grown to the £10,000 mark and the Bristol merchants decided to push ahead with a bridge. They soon discovered that a stone bridge across the gorge would cost more like £90,000, and taking account of recent advances in bridge construction – in particular the use of cast iron as pioneered by the Darbys at Ironbridge, Shropshire, and also Telford's suspension bridge over the Menai Straits – they advertised a competition for the design of a suspension bridge. Brunel immediately threw himself into the task and submitted four of the twenty-two entries received by the closing date. His drawings depicted four designs for bridges at slightly different sites on the gorge, featuring single spans ranging from 760 feet (232 metres) to 1,180 feet (329 metres). This was far in excess of any suspension bridge then in existence. Telford's Menai bridge had a main span of 600 feet (183 metres) while Marc Brunel's suspension bridge on the Ile de Bourbon had two spans, each one just under 132 feet (40 metres). Two of IKB's designs had conventional towers standing on the edge of the cliffs, while the others showed shorter towers with the roadway carved through the rocks and incorporating the existing 'Giants Cave'. With his artistic eye Brunel was keen to capitalise on the drama of the location.

> I thought that the effect... would have formed a work perfectly unique... the grandeur of which would have been consistent with the situation.

Just recently correspondence from 1829 has come to light which shows how much Isambard had sought his father's advice on the design of the bridge. Marc Brunel was concerned that his son's designs for a single span were too ambitious. He wrote to him saying that it 'needed something in the middle to support' and he added, 'you should do it like this', enclosing a sketch with a towering pagoda splitting the gorge and the span into two.

The bridge committee invited the seventy year-old Thomas Telford to judge the entries and he promptly rejected the lot. Instead he submitted his own design and if Brunel had opened a copy of the *Bristol Mercury* in February 1830, he would have been greeted by a depiction of Telford's design for a three span bridge with two lofty Gothic towers rising 260 feet (79 metres) from the shores of the River Avon. It was, proclaimed the newspaper, a 'singularly beautiful design'. As to be expected, Brunel was characteristically candid and sarcastic in expressing his opinion:

> As the distance between the rocks was considerably less than what had always been considered as within the limits of which suspension bridges might be carried, the idea of going to the bottom of such a valley for the purpose of raising expense for two intermediate supporters hardly occurred to me.

He had hit the nail on the head. Telford's towers were simply too expensive and in the end the bridge committee did what committees do best, they fudged it. While publicly expressing their admiration of Telford's Gothic masterpiece, the committee privately reopened discussions and later in 1830 a second competition was announced. Another ragbag of designs was submitted, including some which entirely ignored the stipulation that it was to be a suspension bridge. William Armstrong proposed a single span girder bridge with masonry viaducts to either side, while C. H. Capper put

Thomas Telford didn't believe it was possible to cross the Avon Gorge with a single span and his design, above, featured two Gothic towers rising from the river bank. Brunel's compromise was to reduce the span by creating an abutment on the Leigh Woods side. The Clifton Suspension Bridge was finally opened on 8 December 1864, five years after Brunel's death.

forward a Telford-lookalike with twin towers in a decidedly rustic style. Brunel decided to hedge his bets once again and submitted another four designs. The first was a reworking of the Giant's Cave theme. The second and third reduced the span to 720 feet (219 metres) by introducing a Gothic masonry abutment protruding from the Leigh Woods side, while the fourth pandered to Telford's prejudices by introducing a pair of simplified towers rising from the river bank and finished in an Egyptian style.

Unbelievably, to Brunel at least, his design came in second place in a short-list of four, but all of them came in for criticism from the judges. Brunel was not willing to accept second place and in an audacious display of determination and self-belief he successfully persuaded the committee to accept a modified version of one of his entries. Described by Brunel as the 'Egyptian thing' the approved design featured two towers capped with Sphinxes, with the tower on the Leigh Woods' side resting on an abutment. In June 1831 a ceremony was held to mark the commencement of construction and the foundation stone was laid in August 1836. However, by 1843, with the towers standing ready for the chains, the money had run out and the work was halted.

For two decades the towers loomed over the Avon Gorge like forlorn gravestones to the unfinished bridge. To the Bristolians they became known as 'Brunel's follies'. The chains intended for Clifton were eventually sold off in 1851 to the South Devon Railway Company and they were incorporated within the Royal Albert Bridge at Saltash which was completed shortly before Brunel died in 1859. At the time there were calls for the towers to be removed, but it was IKB's death that prompted his fellow engineers at the Institute of Civil Engineers to have the bridge finished as a fitting monument to their late friend and colleague. John Hawkshaw, the engineer behind the Charing Cross railway bridge being constructed on the piers of Brunel's old Hungerford Bridge, together with W. H. Barlow, wrote a feasibility study showing how the Hungerford chains could be used to complete the Clifton Bridge. There would, however, be several significant changes to Brunel's design. Firstly the Sphinxes and the fine Egyptian decoration on the towers were out. The deck of the bridge was widened from 24 feet (7.3 metres) to 30 feet (9.1 metres) and Barlow added two longitudinal wrought-iron girders to support it. The suspension chains were strengthened with the addition of a third chain and the system of anchoring the chains into the ground was beefed up.

Previous page: In 2006 the Clifton Suspension Bridge was fitted with a bright new lighting system, with 3,000 LEDs plus new floodlights on the towers, cliffs and abutment, to mark the 200th anniversary of Brunel's birth.

Work recommenced on the bridge in 1863 and only eighteen months later, on 8 December 1864, it was officially opened to a cacophony of brass bands, cheering crowds, church bells and the firing of a field gun salute. The bridge that we see now wasn't finished to Brunel's design, but without doubt it is the most famous and fitting memorial to his life and work, and it has become an enduring icon of the city of Bristol.

Above: The bridge photographed from the Rownham Ferry with the river at high tide, *c.* 1910. The Clifton Suspension Bridge Trust was founded by Parliament in 1952 to care for the bridge. One of their first acts was to install thousands of individual light bulbs to commemorate the coronation of Queen Elizabeth II. The lights were hugely popular and the system was upgraded with LEDs and floodlighting in 2006 – *see pages 108 and 109.*

Nothing remains of No. 1 Britain Street in Portsea, Portsmouth, where Brunel was born on 9 April 1806. This plaque marks the approximate spot. The family moved to London when IKB was still an infant and while the family home at Cheyne Walk in Chelsea has survived, Brunel's Duke Street home/office was demolished to make way for new government buildings. Its approximate location is by the entrance to the Cabinet War Rooms, *shown below* on the far right. (*Arpingstone*)

At Home with the Brunels

Events have not been kind to Brunel's houses and only one remains intact.

IKB was born on 9 April 1806 at No.1 Britain Street, in Portsea, Portsmouth – a small house only a vigorous stone's throw away from the navy dockyards where HMS *Victory* now resides and where his father was working on his machinery for manufacturing ship's pulley blocks for the Admiralty. During the Second World War the docks naturally attracted the interest of the Luftwaffe who repeatedly bombed the area. Britain Street, including Brunel's birthplace, and the surrounding area were redeveloped for new housing in the 1960s. Today only a plaque marks the approximate location of the house. To be honest, it's not really worth going out of your way to find it, but if you do there is also a new monument to the Brunels located in the middle of George Street. This was erected in 2006 to mark the 200th anniversary of IKB's birth.

In 1808, Marc Brunel moved his family to London where he invested his money in a sawmill at Battersea. They lived on the opposite side of the river at No. 4 Lindsey Row, now Lindsey House, in Cheyne Walk, a fine Georgian-styled town house which had originally been built in 1674 on the site of Thomas Moore's riverside garden in Chelsea. The house was later separated from the river by the construction of the Chelsea Embankment, which was completed in 1874 to accommodate a new sewage system. The Brunels lived in the centre section of the house, now No. 98, and this is where the young Isambard grew up. According to L. T. C. Rolt it was a very pleasant family home:

'In summer there was swimming in the river from the steps below the house, and exciting excursions into town, then still remote from the rural quiet of Chelsea, made almost invariably by boat.'

In 1820, when he was fourteen years-old, Isambard was packed off to France to study at the College of Caen in Normandy and then in Paris. It was while he was in France that his parents spent a spell in the debtor's prison at Southwark following a calamitous fire at the sawmills. Isambard returned to England in 1822 and he began working in his father's office. With the formation of the Thames Tunnel Company in 1824 Marc Brunel moved family and offices to Bridge Street in Blackfriars. During the construction of the Thames Tunnel, or at least during the phase in which Isambard was involved, father and son had accommodation in neighbouring houses in Cow Court, just to the east of the Rotherhithe site.

Isambard's connection with the tunnel ended with the flooding of the works in January 1828 which so nearly cost him his life. Afterwards he went to Brighton to convalesce and then moved on to Bristol, whether to continue his recovery or because of the work opportunities is not clear. Certainly it was in Bristol that he got his first big breaks with the Clifton Suspension Bridge and the Great Western Railway, but despite his close connections with the city there is no evidence that he ever bought a house in the area. Instead, in 1833, he leased a substantial terraced house at 53 Parliament Street in Westminster, London, to serve as his office. By December 1835, when Brunel was only twenty-nine years old, he moved into a bigger house around the corner at 18 Duke Street, as recorded in his journal on Boxing Day, 1835:

'I am just leaving 53 Parliament Street where I may say I have made my fortune – or rather the foundations of it – and have taken Lord Devon's house, 18 Duke Street. A fine house. I go sometimes with my four horses – I have a cab and a horse – I have a secretary – in fact I am now a *somebody*.'

The new house was an eighteenth-century brick-built four-storey building with an entrance on the Duke Street side and views at the back looking over St James's Park. In addition to providing office space on the lower floors for the engineering business, Brunel wanted to make Duke Street a family home. The following year he proposed to Mary Horsley and they were married in Kensington Parish Church on 5 July 1836. The young bride soon put her mark on the Duke Street house and by all accounts she enjoyed flaunting the refinements that her new wealth and status could bring.

In 1848, Brunel also bought the lease to No. 17, the neighbouring property on the north side, and by the 1851 census the household had expanded to include IKB and Mary, their three children Isambard, Henry Marc and Florence, his widowed mother Lady Sophia Brunel, plus a governess, nurses and various household staff. Brunel's team of engineering assistants also increased substantially as he took on an ever increasing workload and on many of the less important projects his role inevitably became more supervisory. On the long list of minor railways and branch lines much of the detailed work was carried out by his staff, in particular by his Chief Assistant Robert Bereton. Without doubt IKB was a hard taskmaster and a very demanding employer, which was not unusual for the times, and by all accounts his team had nothing but respect and awe for his enormous creativity and endless energy. Likewise his attitude to his family could be decidedly Victorian at times, although he was well loved by his children and he has been described as a 'fond and affectionate father'.

A house in the country

After being appointed as engineer to the Bristol & Exeter Railway and the South Devon Railway, Brunel became enchanted with the West Country and would take his family on annual summer holidays in the area. With the engineering business doing well his thoughts turned to the prospect of retiring near the coast, and in 1847 he purchased an estate at Watcombe, overlooking Babbacombe Bay, about three miles north of Torquay. He immediately set about laying out an elaborate terraced garden and over-seeing the planting of trees with the help of the distinguished landscape gardener William Nesfield, who had previously designed Regent's Park, among other public parks, in London. Brunel's son, another Isambard, later recalled:

'When Mr Brunel bought this property it consisted of fields divided by hedgerows; but, assisted by Mr William Nesfield, he laid it out in plantations of choice trees. The occupation of arranging them gave him unfailing pleasure; and, although he could seldom spare more than a few days' holiday at a time, there can be little doubt that the happiest hours of his life were spent walking about in the gardens with his wife and children, and discussing the condition and prospects of his favourite trees.'

There wasn't a house at Watcombe and in his notebooks Brunel sketched out a design for a house in an Italianate style, although he later commissioned drawings for a grand mansion in a highly decorative French style from the architect William Burn. IKB also designed several cottages and a chapel for the estate workers at Watcombe. Unfortunately the protracted difficulties with the *Great Eastern* became all consuming, both of his time and his money, and work at Watcombe came to a halt. The house was never built and it is probable that Brunel had planned to sell the estate. Such was the severity of his financial situation by 1858 that he had a detailed inventory drawn up of the contents of Duke Street, presumably with a view to selling up if required. Brunel had done well enough financially, but he had an unfortunate habit of investing in his own projects and he often accepted interesting commissions on an unpaid basis. When he died in 1859 his fortune was estimated at £89,000, whereas his close friend and contemporary Robert Stephenson, who died only a month afterwards, left a far more impressive £400,000.

In 1847, Brunel purchased an estate at Watcombe, overlooking Babbacombe Bay near Torquay, in Devon, to create a country home, possibly with a view to his eventual retirement. He played a substantial part in laying out the grounds, aided by the landscape designer William Nesfield, and took great pleasure in planting the trees. However, both time and money ran out before the Brunel's could build a house there. The present house at Watcombe Park is known as Brunel Manor, but its design has no connection with IKB's intended home.

Mary lived in the Duke Street house until her death in 1881, and IKB's former assistant, Robert Bereton, continued to run an engineering consultancy from the premises for several years. Both the Parliament Street and Duke Street houses have since been demolished to make way for new government buildings. In fact, the whole of Duke Street has vanished under the HM Treasury Building on Horse Guards Parade which was completed in 1917. Numbers 17–18 Duke Street were located more or less where the entrance to the Cabinet War Rooms is today. The Watcombe Park estate in Devon was sold and the present house is known as Brunel Manor, although it has nothing to do with any of IKB's designs, and it serves as a Christian holiday and conference centre.

Brunel never got to complete his country home at Watcombe Park. He died on 15 September 1859 at the age of fifty-three.

After an extraordinary transatlantic salvage operation the *Great Britain* returned to Bristol in 1970 to be welcomed by huge crowds as she passed under the Clifton Suspension Bridge. Today she sits in the Great Western Dry Dock – named after the shipping company – where she was originally built and launched in 1843. The ship is now the centrepiece of the revitalised city docks.

Lost and Found

I have to admit that when I started work on *The Lost Works of Isambard Kingdom Brunel* I thought it was going to be a straightforward exercise, a case of amassing material from what I already knew. After all, I reasoned, Brunel and I had an on-going relationship stretching back thirty years to my college days when I made my first photographic survey of his works. But, as it turns out, I couldn't have been more wrong and I have been astounded by the volume of lost works. One reason for this is that I have included some of the un-built works which, in many cases, are represented by one of the hundreds of sketches he made in his notebooks. Then there is the difficulty in making a list of all of Brunel's works, whether lost or not. This is partly because he was so prolific – more on that later – and while the bigger showpiece projects are readily identified, the smaller more mundane works were not necessarily recorded. In some cases original drawings have disappeared over time, especially with the 1,200 miles (1,930 km) or so of railway line to his credit, and in other cases there are no records of 150 years or more of the sort of on-going alterations that can result in the original being buried. This has happened with the railway bridge over the Avon in Bristol. It is a Grade I listed structure, but almost entirely hidden from view by later girder bridges added on either side.

Another difficulty lies in correctly attributing Brunel's work to him. It has often been suggested that Brunel built the modern world, and in his mind this might have been true, but he couldn't possibly have produced such an extensive body of work in his relatively short career without a lot of help from his engineering team. Brunel eschewed the term 'consultant engineer' as it appeared to diminish his starring role as 'Engineer in Chief', but in reality his practice was much like that of one of the better known modern architects for example. So while he had a hands-on approach to the work, admittedly to a lesser or greater degree depending on the project, it was his name, and his name alone, that accrued the glory. However, as we have seen with several projects covered in this book, much of the more laborious work, the leg work of surveying or the time-consuming preparation of countless drawings, was carried out by others under his direction. The Design Museum's 2000 publication, *Isambard Kingdom Brunel – Recent Works*, which accompanied the exhibition of the same name, lists nearly fifty individuals who were either employed by Brunel or worked in collaboration with him. The latter included Matthew Digby Wyatt,

By the mid-1970s, Brunel's Temple Meads station at Bristol was in a sorry state. Cut off from the rails it was being used as a car park, the fine wooden roof neglected and crumbling. Thankfully the 150th anniversary of the GWR in 1985 sparked a restoration programme and the buildings now house a museum and events space. At the other end of the line, Brunel's other great terminus, Paddington, underwent an extensive restoration in the 1990s and is in great shape.

who did the architectural detailing on Paddington Station, and Daniel Gooch, the GWR's first Chief Locomotive Engineer.

None of this is meant to belittle Brunel's incredible achievements, but it is mentioned here to demonstrate how he managed to maintain so many fingers in so many engineering pies. Neither is this account intended as a catalogue of failures as it aims instead to turn the spotlight away from the high-profile works in order that we don't lose sight of the rest. Make no mistake, there is much of Brunel's work that could still be under threat, either through neglect, ignorance or through well-intended improvement. Only recently we very nearly lost the fourth span at Paddington, admittedly a non-Brunellian Edwardian addition to the station, but a timely reminder that the past is not guaranteed protection. Back in the 1950s and 1960s it was all about modernisation and with the Victorian engineers out of fashion several major stations were under threat. Poor old Euston and its iconic arch fell victim to the wreckers, while Victoria Station and Kings Cross were candidates for similar redevelopment. Luckily they survived largely intact, although Victoria Station in particular is suffering from development creep with new buildings pressing in from every side. As a nation we have fallen back in love with Brunel and his ilk and the modern way is for a more sympathetic approach as demonstrated at Kings Cross where the façade is being relieved of its 'modern' clutter, and next door St Pancras has become a beacon of hope. Even so, it has recently been announced that the London to Bristol mainline is to be electrified, and while the travellers might welcome faster trains nothing has been said about the physical or visual impact that the overhead gantries and lines might have on Brunel's great railway.

It's not all doom and gloom. On the plus side many of Brunel's former works have been receiving some TLC recently, and to counterbalance the lost works there are many cases of rediscovered and rescued works. Topping the list is the *Great Britain*, towed back to Bristol in 1970 she now resides in the Great Western Dock where she was built and launched in 1843. After years of painstaking restoration the ship is now the centrepiece of Bristol's revitalised docks, riding on a sea of glass protecting the iron hull from the weather and cocooning it in a desert-dry environment to keep the rust at bay. Bristol is, of course, the city most associated with Brunel and it was also back in the 1970s when I first visited his terminus at Temple Meads, Bristol. It was in a shocking state, cut-off from the rails and left to rot serving out its time as a car park. Thankfully the 150th anniversary of the GWR in 1985 brought its plight to the public's attention and the station building has since been restored to something of its former glory. I only wish the former Engine Shed, immediately behind the GWR offices and now housing the Commonwealth Museum, had retained more of its original identity.

In London, the Brunel Engine House on the Rotherhithe side of the Thames Tunnel has been very imaginatively restored to become the Brunel Museum, and recently the base of the nearby shaft has been sealed to create an exciting new space for future events. At Millwall, on the Isle of Dogs in London, part of the *Great Eastern*'s slipway has been uncovered. Another relic from the ship, part of a funnel, was discovered in Weymouth where it was being used within the town's water supply system, and this is now on display at the *Great Britain*'s museum. Then, in 2004 Dr Steven Brindle of English Heritage located one of IKB's earliest iron bridges hidden from view beneath the brickwork of the later Bishop's Road Bridge where it passed over the Grand Junction Canal at Paddington. The superstructure had long gone, but the iron arches have been rescued and there are plans to reconstruct the bridge nearby as a canal crossing.

Without doubt Brunel's greatest legacy lies in his body of work and, hopefully, this account will help to fill some of the gaps in our appreciation of its breadth and the scope of his genius.

Replica of the *Great Britain*'s original six-bladed propeller, photographed from underneath the hull in the Great Wetern Dockyard where she was built.

At Stroud, Gloucestershire, the broad gauge Goods Shed has survived unscathed apart from a new steel lintel above the doorway, and it is to become a venue for local arts groups. Relics of the *Great Eastern* can be seen at the *Great Britain* museum in Bristol. These include the ship's whistle and a section of funnel, left behind when the ship put into Weymouth for repairs following the explosion, where it remained unnoticed for years as part of the town's water works.

Postscript: A Lost Brunel

There is a curious and rather disturbing footnote to this catalogue of Brunel's lost works, and it involves a disappearing Brunel.

After his death in 1859, the life and work of Isambard Kingdom Brunel was commemorated in a number of ways. He was buried at the Kensal Green Cemetery in London, a sprawling cemetery that might seem an unimposing resting place for Britain's most famous engineer, but this had been a very fashionable location in Victorian times and it is where his parents were already buried. The grave stone is a very simple affair, but as we have seen the Clifton Suspension Bridge was completed as a memorial to Brunel, and on the Royal Albert Bridge at Saltash the epitaph 'I.K. Brunel Engineer' was applied at each end above the portals. Other monuments followed, including a memorial window dedicated to IKB in the nave of Westminster Abbey and, in 1877, after prolonged negotiations regarding its location, Baron Carlo Marochetti's statue of Brunel was finally installed on the Victoria Embankment in London. Here he stands, minus trademark stove-pipe hat and cigar, gazing out across the Thames to where the Hungerford Bridge once stood.

In 1870, Brunel's son, another Isambard, published a very worthy account of his father's career, *The Life of Isambard Kingdom Brunel – Civil Engineer*. However by the early twentieth century the Victorian men of iron were largely out of favour in a world hungry for modernity and it wasn't until Tom Rolt's celebrated biography of Brunel was published in 1957, that a new generation of Brunel's fans began to emerge. Since then his popularity has been on the ascendancy and in 2002, championed by Jeremy Clarkson, he came second in the BBC's poll for the 100 Greatest Britons. In 2006, the bicentenary of his birth was celebrated in grand style, including new lights on the Clifton Suspension Bridge and a spectacular fireworks display.

Nowadays, you are likely to find statues of Brunel scattered all over the place including at Paddington Station, in Swindon, Saltash and, predictably, enough, in Bristol. Inevitably, some of these are better than others – no names mentioned but one high-profile offering has been compared to Charlie Chaplin. One of the finest statues was located at the former Great Western Railway terminus at Neyland in

Opposite: Robert Thomas's statue of Brunel, with broad gauge train in one hand and the *Great Eastern* in the other, was erected at the quayside at Neyland, Pembrokeshire, in 1999. (*Frank Whittle*)

Pembrokeshire. This 8 foot (2.4 metre) high bronze was created by the Welsh sculptor Robert Thomas as a tribute to the town's close links with Brunel who had brought the railway to the area in 1856. Funded with £30,000 of money raised over ten years by the local community, the statue was erected in pride of place on the quayside in 1999. It depicted Brunel holding in one hand a model of a broad gauge train and the *Great Eastern* steamship in the other. Then in August 2010 the shocking news came that this statue, top hat, steamship, loco and all, had been stolen. It had been wrenched from its plinth by thieves and no doubt sold as scrap metal for a fraction of its real value. Sadly, the sculptor, Robert Thomas, has passed away and there is little hope that the people of Neyland will ever get their statue back.

One of the most recent statues of IKB was unveiled at Brunel University in 2006. Sculptor Anthony Stones is shown working on the clay model from which the cast was produced. (*Brunel University*)

The seated figure of Brunel welcomes passengers at Paddington Station, one of two statues by John Doubleday. Baron Marochetti's 1877 statue on the Embankment in London, gazes out towards Hungerford Bridge, while in Swindon a similar bareheaded figure stands above the shoppers on a plinth that resembles a locomotive's chimney.

Further reading

Beckett, Derrick, *Brunel's Britain* (David & Charles, 1988).

Brindle, Steven, *Brunel – The Man Who Built the World* (Weidenfeld & Nicholson, 2005).

Brindle, Steven, *Paddington Station – Its History and Architecture* (English Heritage, 2004).

Brunel, Isambard, *The Life of Isambard Kingdom Brunel – Civil Engineer* (Longmans Green, 1870).

Buchanan, Angus, *Brunel – The Life and Times of Isambard Kingdom Brunel* (Hambledon & London, 2001).

Christopher, John, *Brunel's Kingdom – In the Footsteps of Britain's Greatest Engineer* (Tempus, 2006, History Press, 2009).

Christopher, John, *Isambard Kingdom Brunel Through Time* (Amberley Publishing, 2010).

Christopher, John, *Paddington Station Through Time* (Amberley Publishing, 2010).

Corlett, Ewan, *The Story of Brunel's SS Great Britain – The Iron Ship* (Conway Maritime Press, 2002).

Dugan, Sally, *Men of Iron* (Channel Four Books, 2003).

Hudson, Angie, Eric Kentley and James Peto (eds.), *Isambard Kingdom Brunel – Recent Works* (Design Museum, 2000).

Kelly, Andrew and Melanie (eds.), *Brunel – In Love With the Impossible* (Brunel 200, 2006).

Peak, Alan S., *The Great Western at Swindon Works* (Ian Allan Publishing, 1998).

Rolt, L. T. C., *Isambard Kingdom Brunel* (Longmans Green, 1957).

Vaughan, Adrian, *Isambard Kingdom Brunel – Engineering Knight Errant* (John Murray, 1991).

Acknowledgements

I would like to thank the following individuals and organisations for providing photographs and other images for this book: Arpingstone, Elaine Arthurs of the STEAM Museum of the Great Western Railway, Lance Bellers, Campbell McCutcheon of Amberley Publishing (CMcC), Robert Hulse of the Brunel Museum at Rotherhithe, Hulton Archive/Getty Images, Network Rail, John Powell of the Ironbridge Gorge Museum, the US Library of Congress (LoC), and Frank Whittle. Unless otherwise indicated, all new photography is by the author.

In addition I am grateful to Peter James for helping me to formulate the concept for this book on Brunel's lost works, and to my wife, Ute, for proofreading and patience. JC